Scientific Instruments and Collections

VOLUME 6

The titles published in this series are listed at *brill.com/sico*

Heaven and Earth United

Heaven and Earth United

Instruments in Astrological Contexts

Edited by

Richard Dunn
Silke Ackermann
Giorgio Strano

BRILL

LEIDEN | BOSTON

Cover illustration: Robert Fludd, *Utriusque cosmi maioris scilicet et minoris metaphysica, physica atque technica historia*, Oppenheim, Johann Theodor de Bry, 1617–1621, 2 vols.; v. 2, book I, section II, part IV, p. 71 (Getty Research Institute).

Library of Congress Cataloging in Publication Control Number: 2018027223

Typeface for the Latin, Greek, and Cyrillic scripts: "Brill". See and download: brill.com/brill-typeface.

ISSN 1877-7066
ISBN 978-90-04-36275-8 (hardback)
ISBN 978-90-04-38143-8 (e-book)

This book is printed on acid-free paper and produced in a sustainable manner.

Contents

Figures and Tables

Figures

Tables

Contributors

Silke Ackermann

studied History, Languages and Cultures of the Orient, and History of Science at Frankfurt University (Germany). She worked for sixteen years in a variety of curatorial and managerial roles at the British Museum before taking up the Directorship of the Oxford University Museum of the History of Science following a two-year stint at a German University between these two roles. She served as President of the Scientific Instrument Commission of the International Union for the History and Philosophy of Science until 2017.

Marisa Addomine

is an independent researcher in the history of mechanical horology. Mainly interested in public clocks and early clock-making, in 2004 she discovered a clock in Chioggia which she could date to 1386. She is currently collaborating with Italian and foreign universities and museums, and has been invited to speak at the British Horological Institute and the International Medieval Congress, Leeds.

Jim Bennett

is Keeper Emeritus at the Science Museum, London. He was formerly Director of the Museum of the History of Science and Professor of the History of Science in the University of Oxford. He is a previous President of the Scientific Instrument Commission.

Marvin Bolt

is Curator of Science and Technology at the Corning Museum of Glass. He investigates seventeenth-century refracting telescopes and other optical devices, the role of glass in scientific instruments, and connections between art, science and the humanities.

Louise E. Devoy

is Curator of the Royal Observatory, Greenwich, and has worked in the museum sector for over ten years. She originally studied Physics with Astrophysics at the University of Leicester and later pursued studies in the history of science and museums at Imperial College London. She has since worked at the Science Museum and British Museum, where she discovered an interest in astronomical instruments of the early modern period, which inspired the research featured in this publication.

Richard Dunn
is Senior Curator for the History of Science at Royal Museums Greenwich. His publications include *The Telescope: A Short History* (2009), *Finding Longitude* (2014, with Rebekah Higgitt) and *Navigational Instruments* (2016).

Seb Falk
is a fellow of Girton College, University of Cambridge. He researches late-medieval sciences, especially in religious contexts.

Stephen Johnston
received his BA and PhD from Cambridge University and was a curator at the Science Museum, London from 1987 to 1995. He joined the Museum of the History of Science, University of Oxford in 1995 and is now Head of Research, Teaching and Collections there. His publications have focused on practical mathematics from the sixteenth to the eighteenth centuries and he has curated exhibitions on a broad range of themes within the history of science.

Richard L. Kremer
teaches history of science at Dartmouth College (USA), where he also curates that institution's King Collection of Historic Scientific Instruments. He has published many articles on medieval and early modern astronomical instruments and on the astronomical content of early printed almanacs and calendars. He co-edited *Instruments on Display* (Brill, 2014), volume 4 of the *Scientific Instruments and Collections* series, and currently serves as President of the Scientific Instrument Commission.

Günther Oestmann
trained as a clockmaker. He received a PhD with a study on the astronomical clock of Strasbourg Cathedral in 1992, and in 2001 completed a postdoctoral thesis on Heinrich Rantzau and his attitude towards astrology. He was a lecturer at the Technische Unversität Berlin from 2010 before being appointed as extraordinary professor for the history of science in 2017. In 2013 the Musée international d'horlogerie in La Chaux-de-Fonds awarded him the "Prix Gaïa".

Josefina Rodríguez-Arribas
has research experience in the history of medieval science, Hebrew language and divinatory practices. Since 2009 she has focused on the relations between the textual and material scientific cultures of Jews, notably the cultural role of astrolabes among Jews in Europe and the Near East. She has published on medieval astronomy and astrology and has been a Fellow at Harvard University,

the University of Oxford, the Hebrew University of Jerusalem, the University of Michigan, the University of London and the University of Erlangen-Nuremberg.

Petra G. Schmidl
studied history, history of science and ethnology, and Arabic including supplementary studies in physics, Latin, Hebrew and Persian. She received her doctoral degree from Frankfurt University in 2005 for work on three folk astronomical treatises from eleventh-century Hijaz and thirteenth-century Yemen. Her main research interests lie in the field of pre-modern astronomy and astrology in Islamicate societies, and astronomical instruments, mainly the astrolabe.

Giorgio Strano
PhD, is curator of the collections at the Museo Galileo: Istituto e Museo di Storia della Scienza in Florence. He is involved in the study and popularization of the history of mathematical astronomy. He has published articles in Italian and international journals and has collaborated on the creation of exhibitions on the history of astronomy and science.

Sylvia Sumira
is an independent conservator specializing in globes. After qualifying in paper conservation, she held a Leverhulme studentship in globe conservation at the National Maritime Museum, Greenwich, for several years before setting up her own studio. An accredited member of the UK Institute of Conservation, her clients have included museums, libraries and other institutions in Britain and abroad. Her book *The Art and History of Globes* was published by the British Library in 2014.

Introduction

Historically, the notion that the stars and planets influence the Earth and its inhabitants in some way has been a potent element in almost every culture. What that might mean in practice, however, has been interpreted in countless ways at different times and places, with anything from weak to strong notions of celestial influence invoked within varying intellectual frameworks from the popular to the intensely philosophical. Different interpretations and practices can also be evident even within individual cultural settings. In other words, there is no single thing we can define as astrology, rather a range of astrological contexts, hence the subtitle of this volume.

In the past, historians of astronomy tended to pay little attention to astrology and any related, "non-scientific" topics. In this way, they neglected the fact that until the seventeenth century, astronomy and astrology were so closely interconnected that they almost represented two sides of the same coin. In a similar vein, historians of astrology have often concentrated on the analysis of texts but have neglected to investigate the surviving material culture. Yet countless objects purposefully decorated with astrological imagery survive, as do many instruments, including those within texts, with explicitly astrological purposes or elements. Nevertheless, these artefacts alone cannot give us a full picture of the instrumental practices of different astrological traditions. Printed and manuscript texts discussing instruments and their uses do, of course, offer important contextual evidence, documenting practices that include the political and symbolic alongside the practical. There is also a rich iconography of astrological practice and of celestial influence, some of it instrumental. Together, these different sources offer rich material for investigating the history of astrology (and related practices) and the possible roles of instruments within it. Making a first foray into these possibilities was the intention of a panel organized as part of the XXXIV Scientific Instrument Symposium, held at the Archivio Scientifico e Tecnologico of the University of Turin, Italy, in September 2015. The papers delivered in Turin, with one from a previous Symposium, have been brought together in this volume.

The chapters begin with considerations of the instrument most often linked with astrological practice, the astrolabe. Looking at textual sources, Petra Schmidl considers the earliest surviving evidence on the making and use of astrolabes for astrological purposes. Seb Falk, by contrast, looks to surviving

 | DOI:10.1163/9789004381438_002

instruments and what they can tell us about the effects on usage of the choices made by makers during construction, as well as the effects of subsequent alterations.

The second set of chapters looks at instruments for which a major function was to be an aide-mémoire, a compendium of data to support astrological practices that might incorporate other instruments. Josefina Rodríguez-Arribas looks to medieval and Renaissance instruments for medical uses, highlighting one of the most significant associations of astrological knowledge. Stephen Johnston examines in detail two astrological diptychs made by the English mathematical practitioner, Thomas Hood, in the late-sixteenth century. Coining the useful term "tabular instrument", Günther Oestmann discusses a unique device once owned by Heinrich Rantzau, the Danish governor of Schleswig-Holstein in the second half of the sixteenth century. Louise Devoy looks not to a specific instrument but to the aspectarium, a feature on many early modern instruments. The fact that aspectaria seem not always to have been intended for the determination of astrological aspects is an important conclusion as we think about what an astrological instrument might be.

The next two chapters move from handheld instruments to those placed in public. Marisa Addomine examines some Italian astronomical clocks, thinking about their possible use as astrological machines. Beginning with a German sundial first erected in 1550, the Görlitz Arachne, Richard Kremer then looks at the display of what he terms "public astrological timekeeping".

As instrumental representations of the heavens, globes and armillary instruments have an obvious affinity with astrological conceptions of the universe. The last group of chapters draws on this association. Giorgio Strano describes Antonio Santucci's extraordinary armillaries, which survive in Florence and Madrid, uncovering their construction and potential uses. Jim Bennett and Sylvia Sumira ask whether globes were really used in astrological practice in Early modern Europe, producing interesting evidence and avenues for future research. Approaching from a different direction, Richard Dunn explores the literary and pictorial iconography of the astrologer, who is often associated with a globe, and the extent to which this imagery might have related to quotidian practice.

The final chapter by Marvin Bolt draws together some of the themes and questions raised by the volume's chapters and is framed as a response to them. As he notes, one of the implications of the work presented is that there is still much we do not yet fully understand about how, and indeed how often, instruments were used in different astrological practices. What this volume offers, then, is a first step in the exploration of a rich subject and a challenge for future researchers to explore more fully some of the questions raised and to uncover further evidence of the astrological contexts in which instruments once found a place.

Richard Dunn
Silke Ackermann
Giorgio Strano
April 2018

CHAPTER 1

Using Astrolabes for Astrological Purposes: The Earliest Evidence Revisited

Petra G. Schmidl

In general,* when pre-modern and modern texts and images—and today, even films—depict someone with an astrolabe, it suggests not only that they are practising astronomy, but also that they are familiar with astrology, and perhaps even magic.[1] To what extent, however, was the astrolabe really an astrological instrument?

To begin to address this question, this chapter revisits the earliest evidence concerning astrolabes, texts and instruments, bearing in mind their reliance on earlier traditions.[2] First, it will examine treatises on the construction and use of the astrolabe, two from late Antiquity and the early Byzantine period by Ioannes Philoponus (*c.* 490–*c.* 570) and Severus Sebokht (*c.* 575–666/667), in Greek and Syriac respectively, and two of early ʿAbbāsid times by al-Khwārizmī

* I would like to thank the institutions and individuals who assisted with this chapter, especially the International Consortium for Research in the Humanities "Fate, Freedom and Prognostication. Strategies of Coping with the Future in East Asia and Europe" (supported by the Federal Ministry of Education and Research) at the University of Erlangen-Nuremberg, Germany, as well as Benno van Dalen, Bojidar Dimitrov, Stephan Heilen, David King, Josefina Rodríguez-Arribas and Martin Schmidl for answering questions and sharing their knowledge.

1 For example, *Harry Potter and the Chamber of Secrets*, directed by Chris Columbus, Warner Bros., UK & USA, 2002, in which the office of the headmaster, Albus Dumbledore, is adorned with bookshelves and astrolabes to emphasize that he is a wise and skilled wizard. An earlier example is the story of the barber in *One Thousand and one Nights*, who offers not only to cut hair but also astrological services using an astrolabe; see Enno Littmann (ed.), *Die Erzählungen aus den Tausenundein Nächten*, Frankfurt, Insel Verlag, 2004, 6 vols.; v. 1, pp. 347–349; Juan Vernet, "La conjunción del barbero de Bagdad", in P. Gallais, Y.J. Riou (eds.), *Mélanges offertes à René Crozet*, Poitiers, Société d'études médiévales, 1966, 2 vols.; v. 2, pp. 1173–1175; Jan P. Hogendijk, "A New Look at the Barber's Astrolabe in the *Arabian Nights*", in Jan P. Hogendijk, A. Vrolijk (eds.), *O ye Gentlemen: Arabic Studies on Science and Literary Culture in Honour of Remke Kruk*, Leiden, Brill, 2007, pp. 65–76.

2 See for example Otto Neugebauer, "The Early History of the Astrolabe: Studies in Ancient Astronomy IX", *Isis* 40 (1949), pp. 240–256; David A. King, *In Synchrony with the Heavens: Studies in Astronomical Timekeeping and Instrumentation in Medieval Islamic Civilization*, Leiden, Brill, 2004–2005, 2 vols.; v. 2, pp. 411–415.

 | DOI:10.1163/9789004381438_003

(d. before 850) and ʿAlī b. ʿIsā (fl. *c.* 830) in Arabic.[3] A third text by al-Farghānī (fl. *c.* 900), from a similar context, proves less rewarding. Second, the chapter will investigate three examples of the early ʿAbbāsid astrolabe making tradition: the "Baghdād astrolabe", undated but probably the earliest surviving astrolabe; the earliest dated astrolabe, by Nasṭūlus; and one by al-Khujandī, probably the earliest with astrological markings.

Texts

Ioannes Philoponus and, shortly afterwards, Severus Sebokht, wrote the earliest known treatises on the construction and use of the astrolabe. While Philoponus, a Christian scholar, lived and worked in late fifth- and sixth-century Alexandria in Egypt, Sebokht, a bishop and monk, flourished in late sixth- and seventh-century Asia Minor.[4] Each arranged their treatise in two parts, the first dealing with construction, the second with use.

3 The following editions and translations have been used: Josef Drecker, "Des Johannes Philiponos Schrift über das Astrolab", *Isis* 11 (1928), pp. 15–44; Herbert Wilson Greene, "Treatise Concerning the Using and Arrangement of the Astrolabe and the Things Engraved Upon It: That is to Say, what Each Signifies (by) John Philoponos of Alexandria", in Robert T. Gunther, *The Astrolabes of the World*, Oxford, Oxford University Press, 1932, 2 vols.; v. 1, pp. 61–81; Claude Jarry (ed.), *Jean Philipon: Traité de l'astrolabe: Texte établi et trad.*, Paris, Belles Lettres, 2015; Alfred Stückelberger (ed.), *Ioannes Philoponos: De Usu astrolabii eiusque constructione*, Berlin, Walter de Gruyter, 2015; M.F. Nau, "Le traité sur l'astrolabe plan de Sévère Sabokt", *Journal asiatique* 9 (13) (1899), pp. 56–101 and 238–303; Jessie Payne Margoliouth, "Description of the Astrolabe by Severus Sebokht", in Gunther, op. cit., v. 1, pp. 82–103; Josef Frank, "Die Verwendung des Astrolabs nach al Chwârizmî", *Abhandlungen zur Geschichte der Naturwissenschaften und der Medizin* 3 (1922), pp. 6–32, reprinted in Fuat Sezgin (ed.), *Arabische Instrumente in orientalistischen Studien*, Frankfurt, Institut für Geschichte der arabisch-islamischen Wissenschaften, 6 vols.; v. 4, pp. 130–159, and in Fuat Sezgin (ed.), *Islamic Mathematics and Astronomy*, Frankfurt, Institut für Geschichte der arabisch-islamischen Wissenschaften, 1997–2009, 112 vols.; v. 4, pp. 250–280 and v. 88, pp. 130–159; François Charette, Petra G. Schmidl, "al-Khwārizmī and Practical Astronomy in Ninth-Century Baghdad: The Earliest Extant Corpus of Texts in Arabic on the Astrolabe and Other Portable Instruments", *SCIAMVS* 5 (2004), pp. 101–198; Louis Cheikho (Luwīs Shaykhū), "Kitāb al-ʿamal bi l-aṣṭurlāb li-ʿAlī b. ʿĪsā", *al-Mašriq—al-Machriq. Revue Catholique Orientale mensuelle* 16 (1913), pp. 31–46, reprinted in Sezgin, *Arabische Instrumente in orientalistischen Studien*, cit., v. 3, pp. 368–384 and in Sezgin, *Islamic Mathematics and Astronomy*, cit., v. 87, pp. 366–384; Carl Schoy, "ʿAlī ibn ʿĪsā: Das Astrolab und sein Gebrauch, posthum herausgegeben von J. Drecker", *Isis* 9 (1927), pp. 239–254, reprinted in Sezgin, *Arabische Instrumente in orientalistischen Studien*, cit., v. 3, pp. 385–400, and Sezgin, *Islamic Mathematics and Astronomy*, cit., v. 87, pp. 385–400.

4 Christian Wildberg, "John Philoponos", in Thomas Hockey *et al.* (eds.), *Biographical Encyclopedia of Astronomers*, New York, Springer, 2007, 2 vols.; v. 2, pp. 900a–901b; John M. McMahon, "Severus Sebokht", in *ibid.*, pp. 1044a–1045b.

Neither author mentions astrological markings in the first part of their treatises. For this analysis, "astrological markings" are defined as those that can be used only to solve astrological problems, such as the lines for the twelve astrological houses on standard geographical plates, non-standard plates for casting the rays, and, usually on the back of astrolabes, tables and schemes presenting astrological information relating to the essential dignities.[5] The term "ambiguous markings" refers to those that can be used for solving both astronomical and astrological problems. The line for the horizon on standard geographical plates is a good example, useful for timekeeping and casting horoscopes; similarly, the lines for unequal hours on standard geographical plates, which can be used for timekeeping and to determine the boundaries of the houses.[6] "Non-astrological markings" denote those with no astrological purpose, such as *qibla* quadrants and other nomograms for finding the direction of Mecca, or the lines for prayer times.[7] The following analysis uses the terms "astrological purposes" and "astrological problems" in a similar way.

---

5 The author is not aware of any early Arabic astrolabe that includes a standard geographical plate with additional markings for the houses. An early European example is on an unsigned and undated astrolabe, probably German or Italian, preserved in the Germanisches Nationalmuseum, Nuremberg; see D.A. King, "1.72 Astrolab", in Gerhard Bott (ed.), *Focus Behaim Globus*, Nürnberg, Verlag des Germanischen Nationalmuseums, 1992, 2 vols.; v. 2, pp. 574c–576c. For an example of an early non-standard plate for casting the rays and tables of the dignities, see al-Khujandī's astrolabe; for an overview, see Silke Ackermann, "Astrological Scales on the National Maritime Museum Astrolabes", in Koenraad Van Cleempoel, *Astrolabes at Greenwich: A Catalogue of the Astrolabes in the National Maritime Museum*, Oxford, Oxford University Press, 2005, pp. 73–89.

6 John D. North, *Horoscopes and History*, London, Warburg Institute, 1986, p. 59; Josep Casulleras, Jan P. Hogendijk, "Progressions, Rays and Houses in Medieval Islamic Astrology: A Mathematical Classification", *Suhayl* 11 (2012), pp. 33–102: 83; Josefina Rodríguez-Arribas kindly informed me that she has identified a text on the determination of the astrological houses using horizon, meridian and unequal hour lines on a standard geographical plate, in an anonymous fifteenth-century Hebrew treatise on the astrolabe, Bodleian Library, Oxford, Opp. 697, fol. 19b.

7 An early example of markings on an astrolabe to help find the direction of Mecca is a *qibla* indicator with additional lines for the diurnal prayers and duration of twilight on the back of al-Iṣfahānī's astrolabe (Museo Galileo, Florence, inv. no. 1105), probably made in 1102/1103; see King, *In Synchrony with the Heavens*, cit. (n. 2), v. 2, pp. 528–533. To date the author is not aware of any early eastern Arabic astrolabe that has a standard geographical plate with additional markings for diurnal prayer times. Early western Arabic examples are on three astrolabes made by Ibrāhīm b. Saʿīd al-Sahlī in Toledo and Valencia respectively between June/July 1067 and April 1070. (For Museo Arqueológico Nacional, Madrid, inv. no. 50762, see Gunther, op. cit. (n. 3), pp. 252–253; Salvador García Franco, *Catálogo crítico de astrolabios existentes en España*, Madrid, Instituto Histórico de Marina, 1945, pp. 229–235; J. Carandell *et al.*, *Instrumentos astronomicos en la España medieval: su influencia en Europa*, Santa Cruz de la Palma, Ministerio de Cultura—Dirección General de Bellas Artes y Archivos, 1985, pp. 80–81.

Following these definitions, neither treatise describes astrological markings when dealing with the construction of the astrolabe. The second part of each, on the use of the astrolabe, deals with problems that are ambiguous. For example, both begin with an observation of the solar altitude and the determination of time. Determining time is ambiguous because it is useful for a range of purposes including astrological prognostication.

Two problems discussed in the second parts of the treatises, however, are more obviously astrological. The first deals with the determination of planetary positions.[8] Having set the astrolabe for a specific latitude and time by observing a fixed star, a planet is observed and its position marked on the ecliptic ring of the rete (ignoring its latitude). Noting the positions of the planets is useful for casting a horoscope, but other purposes are conceivable, such as determining rising and setting times and questions of visibility.

The second problem concerns the rising and setting degree of the ecliptic, as well as those on the meridian, for a specific place and time.[9] Knowing the four cardinals aids the drawing of the astrological houses, a division of the ecliptic into twelve parts that depends on the place and time for which they are calculated. They represent the third main element of genethlialogy, besides the zodiacal signs and the planets.[10] Therefore, knowing the four cardinals simplifies casting a horoscope. Determining the ascendant, however, is not useful for astrology alone, but can have other uses, mainly timekeeping.[11] It is revealing

For Museum of the History of Science, Oxford, inv. no. 55331, see *ibid.*, pp. 82–83; Gunther, op. cit. (n. 3), pp. 253–256 and pl. LX. For Osservatorio Astronomico, Rome, inv. no. 157/688, see Gunther, op. cit. (n. 3), p. 263). Only markings purely for observing religious obligations can be classified as non-astrological according to this definition. Even a shadow scale or an hour quadrant is ambiguous since it could aid astrological calculations.

8 Drecker, op. cit. (n. 3), pp. 42–43; Greene, op. cit. (n. 3), v. 1, pp. 79–80; Jarry, op. cit. (n. 3), pp. 40–42; Stückelberger, op. cit. (n. 3), pp. 56–61; Nau, op. cit. (n. 3), pp. 245–247 and 279–280; Margoliouth, op. cit. (n. 3), v. 1, p. 91.

9 Drecker, op. cit. (n. 3), pp. 31–32; Greene, op. cit. (n. 3), v. 1, pp. 70–71; Jarry, op. cit. (n. 3), pp. 21–22; Stückelberger, op. cit. (n. 3), pp. 28–31; Nau, op. cit. (n. 3), pp. 251–254 and 284–287; Margoliouth, op. cit. (n. 3), v. 1, pp. 93–95.

10 See Charles Burnett, "Astrology", in Kate Fleet *et al.* (eds.), *Encyclopaedia of Islam, THREE*, Brill Online, 2015, http://referenceworks.brillonline.com/entries/encyclopaedia-of-islam-3/astrology-COM_0162 (accessed 12 May 2015).

11 "Oblique ascensions are of singular importance in timekeeping and in mathematical astrology […]. In order to measure the time of day, it is thus sufficient to control the oblique ascensions of the horoscopus and the solar longitude": D.A. King, "Maṭāliʿ" in P.J. Bearman *et al.* (eds.), *The Encyclopedia of Islam*, second edition, Leiden, Brill, 1960–2005, 12 vols; v. 6, p. 793b; "The horoscopus is of prime importance in astrology as well as in timekeeping": King, *In Synchrony with the Heavens*, cit. (n. 2), v. 1, p. 75.

that al-Khwārizmī includes problems that use the ascendant as a parameter for timekeeping and the determination of solar and stellar altitudes (see below).

The terminology that Philoponus and Sebokht use to describe this problem may provide further hints. Do they lean towards astrological prognostication or astronomical calculation? Philoponus uses ambiguous terms such as κέντρα, ὡροσκοποῦν and μεσουρανοῦν when discussing the determination of the four cardinals.[12] Most translations echo this, speaking of "centres", "centre du levant" and "milieu du ciel" (Jarry), "Kardinalpunkte", "Horoskop-Punkt" and "Mittagspunkt" (Stückelberger), "Central Points", "Ascendant and Culminating" (Green), and "Spitzen", "Horoskopos" and "Himmelsmitte" (Drecker; see *Appendix 2*). By contrast, Sebokht uses terms clearly related to astrology, as well as ambiguous terms; for instance, terms translated as "milieu du ciel" (Nau) and "middle of the heaven" (Margoliouth).[13] In particular, the ascendant, descendant and lower mid-heaven are described in terms whose translations from the Syriac include "degré 'de vie'", "degré 'des noces'" and "centre 'des pères'" (Nau), i.e. "degree 'of life'", "degree 'of nuptials'" and "centre of 'the fathers'" (Margoliouth; see *Appendix 2*). These terms reflect the indications of the astrological houses (*Table 1.1*), as given, for example, by Julius Firmicus Maternus in the fourth century.

In the *Matheseos*, Maternus refers to the first house as *horoscopus* and relates it to life and soul, the fourth as *imum coeli* and to parents and parental affairs, and the seventh to wives.[14] Both Nau and Margoliouth put the indication of

12 Drecker, op. cit. (n. 3), p. 31, n. 13, argues that he translated κέντρα as cusps and explains that what is meant are the four main points of the astrolabe later called *cardines* or *anguli*. Further he states that Philoponus avoids astrological topics ("Ich habe κέντρα mit Spitzen übersetzt, […] gemeint sind die 4 Hauptpunkte des Astrolabs, die man später *cardines* oder *anguli* nannte. […] Philoponos hält sich vom Astrologischen frei." Since κέντρα is used as a technical term in Greek astrological treatises to describe the four cardinals in a horoscope, Drecker's interpretation falls short (e.g. see Auguste Bouché-Leclercq, *L'astrologie grecque*, Paris, Leroux, 1899, p. 258 and n. 4; Stephan Heilen, *"Hadriani genitura": Die astrologischen Fragmente des Antigonos von Nikaia*, Texte und Kommentare 43, Berlin, Walter de Gruyter, 2015, pp. 732–733; references kindly provided by Stephan Heilen); Jarry, op. cit. (n. 3), p. LXI, questions whether something astrological remains unsaid in this approach and whether Philoponus, without saying so, pre-empts the concerns of his audience; this seems possible to him ("Y a-t-il un non-dit astrologique dans cette démarche, et Philopon va-t-il sans le dire au devant des préoccupations de son public? Cela ne nous semble pas impossible.").

13 Neugebauer, op. cit. (n. 2), p. 245: "But Sebokht repeats the same thing [the determination of the rising times of zodiacal signs] in chapter X, using astrological terminology […]".

14 "Primus est locus [idest] illa pars, in qua horoscopus est constitutus. In hoc loco vita hominum et spiritus continetur […]. Quartus ab horoscopo locus id est IMC. in quarto ab horoscopo signo constituitur […]. Hic locus ostendit nobis parentes patrimonium

TABLE 1.1 The astrological houses and their indications

House	Ecliptic point	Indication
I	ascendant	life
II	–	wealth
III	–	brothers
IV	lower mid-heaven	parents
V	–	children
VI	-	health
VII	descendant	marriage
VIII	–	death
IX	–	journeys
X	upper mid-heaven	power
XI	–	friends
XII	–	enemies

the houses, ("de vie"/"of life", etc.) in inverted commas. Presumably, both read them as technical terms close to the Syriac text. Thus the problem of determining the four cardinals has an astrological colouring, more intense in Sebokht's treatise than in Philoponus', although it remains an ambiguous problem.

There is, therefore, no clear evidence that the main purpose of the astrolabe as described by Philoponus and Sebokht is astrological. When dealing with construction, neither mentions astrological markings. Concerning its use, most chapters deal with timekeeping problems, which are ambiguous. The terminology used, particularly in the chapters on the determination of the four cardinals, however, emphasizes Sebokht's familiarity with astrology and moves this problem closer to astrological usage.

Next in chronological order are the treatises by Muḥammad b. Mūsā al-Khwārizmī, the famous mathematician and astronomer, and his younger colleague and contemporary, ʿAlī b. ʿIsā al-Asṭurlābī, an astronomer and instrument maker, both associated with the court of the ʿAbbāsid caliph al-Maʾmūn

substantiam fundamenta mobilia, et quicquid ad latentes et repositas patrimonii pertinet facultates. [...] Septimus locus in septimo ab horoscopo signo constituitur [...]. Ex hoc loco qualitatem et quantitatem quaeramus nuptiarum": Julius Firmicus Maternus, *Matheseos libri VIII*, ed. by Wilhelm Kroll and Franz Skutsch, Leipzig, Teubner, 1897, 2 vols; v. 1, pp. 61–63.

(d. 833).[15] A third treatise from a similar context, by al-Farghānī, only deals with the construction and theory of the astrolabe, providing little help in addressing the question of the astrological uses.[16]

Al-Khwārizmī's text, which is presumably the older, is more elaborate and detailed.[17] Like Philoponus and Sebokht, he deals in the first part with the construction of the astrolabe, while ʿAlī b. ʿIsā describes only its parts. Neither Arabic author nor al-Farghānī mentions astrological markings.[18]

In the second part on the use of the astrolabe, al-Khwārizmī and ʿAlī b. ʿIsā start, like Philoponus and Sebokht, with measuring solar altitude and continue with ambiguous problems, including the two mentioned above that have an astrological colouring.[19] ʿAlī b. ʿĪsā disregards the first problem, determining the position of the planets, but al-Khwārizmī provides an approximate method that takes both longitudes and latitudes into account.[20]

The second problem, determining the four cardinals, is only found in ʿAlī b. ʿĪsā's text, even though his treatise is much shorter than al-Khwārizmī's.[21] The latter does not discuss the four cardinals in a separate chapter, but provides the determination of the ascendant by different means and makes use of it as a parameter for timekeeping and determining solar and stellar altitudes (*Table 1.2*). This supports the identification of the determination of the four cardinals in the treatises by Philoponus and Sebokht as an ambiguous problem. More telling indicators, however, are the determination of the descendant and upper and lower mid-heaven, since the ascendant alone is sufficient for the non-astrological uses described by al-Khwārizmī. It is also worth noting

15 Sonja Brentjes, "Khwārizmī: Muḥammad ibn Mūsā al-Khwārizmī", in Hockey *et al.*, op. cit. (n. 4), v. 1, pp. 631b–633b; Marvin Bolt, "ʿAlī ibn ʿĪsā al-Asṭurlābī", in *ibid.*, pp. 34a–34b; King, *In Synchrony with the Heavens*, cit. (n. 2), v. 2, p. 456 and n. 47, queries whether ʿAlī b. ʿIsā al-Asṭurlābī's treatise was written in ninth-century Baghdād.

16 Gregg DeYoung, "Farghānī: Abū al-ʿAbbās Aḥmad ibn Muḥammad ibn Kathīr al-Farghānī", in Hockey *et al.*, op. cit. (n. 4), v. 1, p. 357a–357b.

17 Charette, Schmidl, op. cit. (n. 3), p. 166, n. 76.

18 *Ibid.*, pp. 110–114 and 135–139; Cheikho, op. cit. (n. 3), pp. 32–34; Schoy, op. cit. (n. 3), pp. 240–241; Richard P. Lorch, *al-Farghānī: On the Astrolabe*, Stuttgart, Franz Steiner, 2005, esp. pp. 6–7; al-Farghānī, however, discusses the lines for afternoon prayer (Benno van Dalen, pers. comm., 11 December 2015).

19 Frank, op. cit. (n. 3), pp. 6–17; Charette, Schmidl, op. cit. (n. 3), pp. 115–124 and 140–150; Cheikho, op. cit. (n. 3), pp. 34–46; Schoy, op. cit. (n. 3), pp. 241–253.

20 Frank, op. cit. (n. 3), p. 9; Charette, Schmidl, op. cit. (n. 3), pp. 117–118 and 142–143.

21 The title of the chapter reads: "استخراج الطالع والغارب وجزء وسط السماء ووتد الأرض وخطّهما" (Determination of the ascendant and the descendant, the degree of upper and lower mid-heaven and both their lines), Cheikho, op. cit. (n. 3), p. 45; Schoy, op. cit. (n. 3), pp. 242–243, translates it as "Ermittlung des Ascendenten, des Descendenten, des Grades der Mitte des Himmels und des Drehpunktes der Erde und der Linie beider".

TABLE 1.2 Selected chapters of al-Khwārizmī's treatise on the astrolabe

Problem	Content	Frank	Charette and Schmidl
Timekeeping	Ascendant, hours of daylight and their fractions determined from solar altitude	Ch. 2, p. 6	Ch. 2a, pp. 115, 140
	Solar altitude and hours of daylight determined from the ascendant	Chs. 23–24, p. 11	Chs. 17–18, pp. 119–120, 145
	Ascendant and hours of night determined by measuring stellar altitude	Ch. 3, pp. 7–8	Ch. 3, pp. 117, 141
	Stellar altitude determined from the ascendant	Ch. 26, p. 12	Ch. 20, pp. 120, 145–146
Astrology	Astrological houses	Ch. 31, p. 13	Ch. 23, pp. 120–121, 146–147
	Projection of the rays	Ch. 32, p. 13	Ch. 24, pp. 121, 147
	Year-transfer of the nativities	Ch. 33, pp. 13–14	Ch. 25, pp. 121, 147
		Ch. 36b, pp. 14–15	Ch. 26a–c, pp. 121–122, 147–148
Religious performances related to religious observances	Determining the beginning and end of midday and afternoon prayer	Ch. 36a, p. 14	Ch. 28c–e, pp. 122, 148
	Determining daybreak and nightfall (related to evening, night and morning prayer)	Ch. 35, p. 14	Ch. 28a–b, pp. 122, 148
	Determining the time of moonrise (beginning of a month)	Ch. 38, pp. 15–16	Ch. 30a, pp. 123, 149–150
	Determining the azimuth (direction of Mecca)	Ch. 37, p. 15	Ch. 29, pp. 123, 148–149

that al-Khwārizmī's treatise shows evidence of interest in religious obligations and their performances. Thus he includes chapters on the determination of the daily ritual prayers in Islam and the lunar calendar (*Table 1.2*).[22]

22 For an overview, see D.A. King, "Science in the Service of Religion: The Case of Islam", *Impact of Science on Society* (UNESCO) 159 (1990), pp. 245–262, reprinted in D.A. King, *Astronomy in the Service of Islam*, London, Variorum, 1993, art. I. al-Khwārizmī's interest in

Beyond this, what really makes the difference between the first two treatises discussed and those of early ʿAbbāsid times is that both the authors writing in Arabic include chapters of astrological content. Al-Khwārizmī deals with the determination of the astrological houses, the projection of rays and the year-transfer of nativities (*Table 1.2*).[23] ʿAlī b. ʿĪsā's shorter treatise only includes the determination of the houses.[24] Taking together the findings from the earliest Greek, Syriac and Arabic texts on the astrolabe, this emergence of chapters on astrological problems in the treatises by al-Khwārizmī and ʿAlī b. ʿĪsā is, therefore, a conspicuous development.

Instruments

The analysis of surviving instruments provides further evidence. Around a dozen astrolabes made in the eighth to tenth centuries in Baghdād survive; other early examples are mentioned in written sources.[25] Three are considered here.

The first is a tiny astrolabe, probably a kind of instruments known to and used by scholars in eighth- and ninth-century Baghdād. Formerly in the Archaeological Museum, Baghdād, its recent whereabouts are unknown. Although undated and unsigned, the general appearance and design point towards construction in the early ninth or even late eighth century. It has a rete with a small number of star pointers, geographical plates without azimuth curves, and two altitude scales in the upper two quadrants on an otherwise blank back. There are no specific astrological markings, either on the plate or

---

topics as astrological problems and ritual performances is reflected in two of the sources added to the edition and translation of al-Khwārizmī's treatise on the construction and use of the astrolabe: "On a compass for finding the prayer times" and "On projecting the rays with the astrolabe"; Charette, Schmidl, op. cit. (n. 3), pp. 124–126 and 150–152, and pp. 128 and 153–154.

23 Casulleras, Hogendijk, op. cit. (n. 6), pp. 33–102; Edward S. Kennedy, "The Astrological Houses as Defined by Medieval Islamic Astronomers", in Josep Casulleras, Julio Samsó (eds.), *From Baghdad to Barcelona: Studies in the Islamic Exact Sciences in Honour of Prof. Juan Vernet*, Barcelona, Instituto "Millás Vallicrosa" de Historia de la Ciencia Arabe, 1996, 2 vols.; v. 2. pp. 535–578; John D. North, "A Reply to Prof. E.S. Kennedy", in *ibid.*, v. 2, pp. 579–582.

24 Cheikho, op. cit. (n. 3), pp. 39–40; Schoy, op. cit. (n. 3), pp. 247–248.

25 See King, *In Synchrony with the Heavens*, cit. (n. 2), v. 2, pp. 401–437 and 439–544; D.A. King, "An Instrument of Mass Calculation made by Nasṭūlus in Baghdad ca. 900", *Suhayl* 8 (2008), pp. 93–119; D.A. King, "Two newly-discovered astrolabes from ʿAbbasid Baghdad", *Suhayl* 11 (2012), pp. 103–116.

on the back.[26] That said, operations such as the determination of the positions of the planets or the cardinals, as described in texts, can be performed with this or any other astrolabe. Such operations would, however, be difficult with the "Baghdād astrolabe", because of its size, only about 85 mm in diameter. As it is neither dated nor signed and its whereabouts are uncertain, it is a rather weak witness. The same is true for several other astrolabes that have survived from Baghdād in early ʿAbbāsid times. They are also either undated or unsigned or both.

A second astrolabe, from the beginning of the tenth century, bears its maker's name and production date. An inscription on the back of the throne translates, "Nasṭūlus [or Basṭūlus] made it in the year 315 [927–928]". Like the "Baghdād astrolabe", the rete has seventeen star pointers and the plates lack azimuth curves. As well as two altitude scales in the upper two quadrants, the back has another scale in the lower right quadrant for determining the cotangent from an angle (or vice versa). Again, neither the plates nor the back have astrological markings. Since Nasṭūlus' astrolabe is approximately twice the size of the "Baghdād astrolabe", its diameter comprises 173 mm, operations including those for astrological problems could be carried out more easily. Three other astronomical instruments bearing an inscription "made by Nasṭūlus" survive, although none is dated: a mater with a gazetteer; a second complete astrolabe with two plates; and a circular instrument mainly for timekeeping. None of Nasṭūlus' instruments includes astrological markings.[27]

Nearly forty years later, al-Khujandī (fl. second half of the tenth century) made the third example. Its inscription translates, "Made by Ḥāmid b. al-Khiḍr al-Khujandī in the year 374 *hijra* [984–985]". Sources describe al-Khujandī as a skilful mathematician, astronomer and instrument maker, most famous for the enormous mural sextant he used in Rayy for solar observations and measuring the obliquity of the ecliptic, although he also made another handheld device, *al-āla al-shāmila* (the complete instrument).[28] Although it is slightly smaller (153 mm diameter) than Nasṭūlus' astrolabe, its rete has nearly twice the number of star pointers (33). The mater holds two non-standard plates. One is purely for astrological use, in that it displays curves for projecting the

26 For a description, see King, *In Synchrony with the Heavens*, cit. (n. 2), v. 2, pp. 411–433.

27 Mater with gazetteer: King, *Ibid.*, pp. 476–482; complete astrolabe: King, "An Instrument of Mass Calculation", cit. (n. 25), pp. 99–103; circular instrument: King, "Two Newly-discovered Astrolabes", cit. (n. 25), pp. 104–108.

28 Glen van Brummelen, "al-Khujandī: Abū Maḥmūd Ḥāmid ibn al-Khiḍr al-Khujandī", in Hockey *et al.*, op. cit. (n. 4), v. 1, pp. 630b–631b.

TABLE 1.3 Final-letter abbreviations of planets' names

Saturn	*zuḥal*	*l*
Jupiter	*al-mushtarī*	*ī*
Mars	*al-mirrīkh*	*kh*
Sun	*al-shams*	*s*
Venus	*al-zuhra*	*a*
Mercury	*ʿuṭārid*	*d*
Moon	*al-qamar*	*r*

rays, a complex procedure.[29] The other non-standard plate bears ambiguous markings, though with an astrological tendency. It has half arcs of horizons for twenty geographical latitudes from 15° to 55°. It therefore facilitates the determination of the ascendant for any location between these limits, for which the standard plates are not suited.[30] On the back of the astrolabe, al-Khujandī also added tables for the essential dignities, the limits (or terms), faces (also called decans) and the lords of the triplicities (see *Appendix 1*).[31] The way in which al-Khujandī presents the planetary names in the ecliptic tables on the back of the astrolabe is worth noting, since he combines final-letter abbreviations (i.e. the final letter of a planet's name; *Table 1.3*) and planetary symbols. The entries for the limits and the lords of the triplicities use the former, those for the faces use the latter.

Final-letter abbreviation was more common than symbols for denoting planets in pre-modern astronomy, astrology and instrument making. Al-Bīrūnī (b. 973) in his *Kitāb al-Tafhīm fī ṣināʿat al-tanjīm* limits the use of symbols to astrological scales on the back of astrolabes.[32] The same is true for surviving

29 Casulleras, Hogendijk, op. cit. (n. 6), pp. 50–51; King, *In Synchrony with the Heavens*, cit. (n. 2), v. 2, p. 514: "It is the first attestation of such plates, which were used into the Renaissance."

30 *Ibid.* Another early example is on one of the two astrolabes by Ḥāmid b. ʿAlī al-Wāsiṭī; on the back of the second, he uses planetary symbols for astrological tables; *ibid.*, pp. 496–503. See also *Appendix 3*.

31 Willy Hartner, "The Principle and Use of the Astrolabe", in Arthur Upham Pope (ed.), *Survey of Persian Art*, London and New York, Oxford University Press, 1938–1958, 6 vols.; v. 3, pp. 2530–2554 and 2548–2549; reprinted in Günther Kerstein *et al.*, *Willy Hartner: Oriens-Occidens*, Hildesheim, Georg Olms, 1968, pp. 287–311; reprinted in Sezgin, *Islamic Mathematics and Astronomy*, cit. (n. 3), v. 94, pp. 338–375, and as Astrolabica no. 1 (1978); Ackermann, op. cit. (n. 5), pp. 80–81.

32 R. Ramsey Wright, *The Book of Instruction in the Art of Astrology of al-Bīrūnī*, London, Lizac & Co., 1934, p. 199, §329.

astrolabes (*Appendix 3*); other examples in texts are only found in astrological and magical treatises.[33] In using the planetary symbols to present the planets in the ecliptic table on the back of the astrolabe, al-Khujandī seems to highlight their astrological nature and the astrological use of the astrolabe. Thus while al-Khujandī's non-standard plates, one for the projection of the rays, the other for the horizons, facilitate the use of the astrolabe for astrological problems, his astrological tables dispense with the need to look up this information in a book.

Unlike al-Khwārizmī, whose treatise also discusses problems relating to religious obligations, al-Khujandī dos not include solely religious markings on his astrolabe. For the eastern tradition of astrolabe making, they seem only to appear a century later on an astrolabe made by Muḥammad b. Abī l-Qāsim al-Iṣfahānī al-Ṣāliḥānī. One must, however, be wary of comparing the incomparable by juxtaposing al-Khwārizmī's problems on the use of the astrolabe with al-Khujandī's markings as part of its construction.

Conclusion: Using Astrolabes for Astrological Purposes?

Astrolabes seem, therefore, to have had astrological uses in early ninth-century Iraq, in al-Khwārizmī's treatise on the construction and use of the astrolabe, and in late tenth century Iran, on al-Khujandī's astrolabe. The local and temporal shift between these two may be due to the fact that, apart from the vicissitudes of life, astrological problems may not always be perceptible on an astrolabe.

What happened, then, between the fifth and ninth centuries? Can we surmise an increasing interest in astrology over this period? Of course, the answer is not that simple. First, the absence of descriptions of astrological markings and problems does not prove that Philoponus and Sebokht ignored astrological purposes. Although not clearly described, an interest in astrology is perceptible in their treatises, hidden in ambiguous markings and problems.

Second, al-Khwārizmī's, ʿAlī b. ʿĪsā's and al-Khujandī's interest in astrology may not just derive from the subject itself. Over such a large area and long period of time, many things changed that are not that stated in the sources. One

33 P.G. Schmidl, "Magic and Medicine in a 13th c. Treatise on the Science of the Stars: *The Kitāb al-Tabṣira fī ʿilm al-nujūm* of the Rasulid Sultan al-Ashraf ʿUmar", in Ingrid Hehmeyer, Anne Regourd, Hanne Schönig (eds.), *Herbal Medicine in Yemen: Traditional Knowledge and Practice, and their Value for Today's World*, Leiden, Brill, 2012, pp. 43–68: 50–52.

of these may concern the purpose of the astrolabe. Being multi-functional, the astrolabe is adaptable. During this time, one function may have become more important, another less. Perhaps, for example, the representative or the astrological function became more important. Such changes related to the people who wrote on astrolabes and made them as well as to those for whom they were written and produced. Did technical skills change, for example, so that making and using plates for casting the rays became possible? Did social and economic circumstances change so that authors had to cover specific topics in detail, such as problems relating to astrology or religious obligations? Did aesthetic sensibilities change so that blank backs of astrolabes came to be regarded as less beautiful than inscribed ones? These are only a few possibilities that may have led to the inclusion of astrological problems in the later texts.

Conspicuously, problems relating to religious obligations emerged alongside astrological ones in astrolabe treatises in the early ʿAbbāsid period. This supports an explanatory model in which Greek knowledge was appropriated and naturalized in early Muslim times.[34] As a part of this process, sources begin to exhibit a need for universality and pervasiveness as well as an interest in correcting deficiencies and providing gimmicks.[35] Al-Farghānī's book on the astrolabe is just one example that illustrates this. It has several titles, one of them *al-Kāmil fī l-asṭurlāb* (The Complete [Book] on the astrolabe), and includes not only the northern but also the southern astrolabe,[36] in a similar way to al-Khujandī's handheld device, *al-āla al-shāmila*. These assumed

34 Abdelhamid I. Sabra, "The Appropriation and Subsequent Naturalization of Greek Science in Medieval Islam: A Preliminary Statement", *History of Science* 25 (1987), pp. 223–243. For the broader context, see Sonja Brentjes, Robert G. Morrison, "The Sciences in Islamic societies (750–1800)", in Robert Irwin (ed.), *The New Cambridge History of Islam*, vol. 4: "Islamic Cultures and Societies to the End of the Eighteenth Century", Cambridge, Cambridge University Press, 2000, pp. 564–639; Dimitri Gutas, *Greek Thought—Arabic Culture: The Graeco-Arabic Translation Movement in Baghdad and Early ʿAbbāsid Society (2nd–4th/8th–10th centuries)*, London and New York, Routledge & Kegan Paul, 1998, esp. pp. 108–110.

35 D.A. King, "Universal Solutions in Islamic Astronomy", in J. Lennart Berggren, Bernard R. Goldstein (eds.), *From Ancient Omens to Statistical Mechanics: Essays on the Exact Sciences Presented to Asgar Aaboe*, Copenhagen, University Library, 1987, pp. 121–132: 121–123; D.A. King, "Universal Solutions to Problems of Spherical Astronomy from Mamluk Egypt and Syria", in Farhad Kazemi, R.D. McChesney (eds.), *A Way Prepared: Essays on Islamic Culture in Honor of Richard Bayly Winder*, New York and London, New York University Press, 1988, pp. 153–184: 155–156.

36 A.I. Sabra, "Faghrānī, Abu'l -ʿAbbās Ibn Muḥammad Ibn Kathīr al-", in Charles C. Gillespie (ed.), *Dictionary of Scientific Biography*, New York, Scribner's, 1970–1980, 16 vols.; v. 4, pp. 541–545; information kindly provided by Benno van Dalen.

needs and supposed interests could only be pursued in a supportive environment. In early ʿAbbāsid times, courtly patronage helped to promote such interests.[37]

These conclusions remain somewhat speculative. Nevertheless, what remains is persuasive evidence from texts describing the astrolabe first and foremost as a timekeeping device, but suggesting a use for astrological purposes that became more and more explicit, and instruments that accrued astrological markings over time.

Appendix 1: The Astrological Tables on al-Khujandī's Astrolabe

On the lower half of the back of the astrolabe, al-Khujandī provides a semicircular ecliptic table. From the rim to the centre, it lists correspondences of lunar mansions, zodiacal signs, limits or terms, faces and the lords of the triplicities. A possible source is *Mukhtaṣar al-Mudkhal* (The Abbreviation of the Introduction) by Abū Maʿshar (Baghdād, ninth century), or a text closely related to it,[38] since both Abū Maʿshar and al-Khujandī identify Mercury as an additional companion to Mars in Virgo, while, for example, al-Bīrūnī, Kūshyār b. Labbān (fl. around 1000) or al-Qabīṣī (fl. second half of the tenth century) do not in their introductions to astrology.[39]

In the following transliteration, al-Khujandī's table is split into four subtables. For the entries of the limits and lords of the triplicities, al-Khujandī abbreviates the names of the planets by using their final letter. In the table of the faces, he uses planetary symbols.

Entries indicated with a dash are illegible. Numbers in square brackets indicate Abū Maʿshar's variants.

37 Brentjes, Morrison, op. cit. (n. 34), pp. 572–573.

38 Charles Burnett, Keiji Yamamoto, Michio Yano (eds.), *Abū Maʿshar: The Abbreviation of the Introduction to Astrology together with the Medieval Latin Translation of Adelard of Bath Edited and Translated*, Leiden, Brill, 1994, pp. 88–89 (terms), 14–23 (decans) and 24–27 (lords of the triplicities).

39 *Ibid.*, p. 25: "Taurus, Virgo and Capricorn [...] and the participant with them both by day and night is Mars, and Mercury shares with them both in Virgo". Wright, op. cit. (n. 34), p. 259, §445; Michio Yano, *Kūshyār Ibn Labbān's Introduction to Astrology*, Tokyo, Institute for the Study of Languages and Cultures of Asia and Africa, 1997, pp. 36–37; C. Burnett, K. Yamamoto, M. Yano, *al-Qabīṣī (Alcabitius): The Introduction to Astrology: Edition of the Arabic and Latin Texts and an English Translation*, London, Warburg Institute, 2004, pp. 24–25.

al-sharaṭayn	*al-buṭayn*	*al-thurayyā*	*al-dabarān*	*al-haqʿa*	*al-hanʿa*	*al-dhirāʿ*

Aries					Taurus					Gemini				
Jup.	Ven.	Mer.	Mar.	Sat.	Ven.	Mer.	Jup.	Sat.	Mar.	Mer.	Jup.	Ven.	Mar.	Sat.
6	6	8	5	5	8	6	7	5	4[5]	6	6	5	7	6

Mars	Sun	Venus	Mercury	Moon	Saturn	Jupiter	Mars	Sun
10	20	30	10	20	30	10	20	30
Sun	Jupiter	Saturn	Venus	Moon	Mars	Saturn	Mercury	Jupiter
Jupiter	Sun		Moon	Venus		Mercury	Saturn	

al-nathra	*al-ṭarfa*	*al-jabha*	*al-zubra*	*al-ṣarfa*	*al-ʿawwāʾ*	*al-simāk*

Cancer					Leo					Virgo				
Mar.	Ven.	Mer.	Jup.	Sat.	Jup.	Ven.	Sat.	Mer.	Mar.	Mer.	Ven.	Jup.	Mar.	Sat.
6[7]	6	6	7	5[4]	6	5	7	6	6	7	–[10]	–[4]	7	2

Venus	Mercury	Moon	Saturn	Jupiter	Mars	Sun	Venus	Mercury
10	20	30	10	20	30	10	20	30
Venus	Mars	Moon	Sun	Jupiter	Saturn	Venus	Moon	Mercury
Mars	Venus		Jupiter	Sun		Moon	Venus	Mars

al-ghafr	*al-zubānā*	*al-iklīl*	*al-qalb*	*al-shawla*	*al-naʿāʾim*	*al-balda*

Libra					Scorpio					Sagittarius				
Sat.	Mer.	Jup.	Ven.	Mar.	Mar.	Ven.	Mer.	Jup.	Sat.	Jup.	Ven.	Mer.	Sat.	Mar.
6	8	7	7	2	7	4	8	5	6	12	5	4	5	4

Moon	Saturn	Jupiter	Mars	Sun	Venus	Mercury	Moon	Saturn
10	20	30	10	20	30	10	20	30
Saturn	Mercury	Jupiter	Venus	Mars	Saturn	Sun	Jupiter	Saturn
Mercury	Saturn		Mars	Venus		Jupiter	Sun	

al-dhābiḥ	*saʿd bulʿa*	*al-suʿūd*	*al-akhbiya*	*fargh al-muqaddim*	*al-muʾakhkhir*	*baṭn al-ḥūt*

Capricorn					Aquarius					Pisces				
Mer.	Jup.	Ven.	Sat.	Mar.	Mer.	Ven.	Jup.	Mar.	Sat.	Ven.	Jup.	Mer.	Mar.	Sat.
7	7	8	4	4	7	6	7	5	5	12	4	4[3]	9	2

Jupiter	Mars	Sun	Venus	Mercury	Moon	Saturn	Jupiter	Mars
10	20	30	10	20	30	10	20	30
Venus	Moon	Mars	Saturn	Mercury	Jupiter	Venus	Mars	Moon
Moon	Venus		Mercury	Saturn		Mars	Venus	

Appendix 2: Ioannes Philoponus' and Severus Sebokht's Terms for the Four Cardinals

The terms used in Ioannes Philoponus' and Severus Sebokht's treatises may be an indicator of whether the authors thought of astrological prognostication or astronomical calculation when writing about the determination of the cardinals. The following table lists the translations of terms used for the cardinals in both treatises.

Modern designation	**Ioannes Philoponus** **Stückelberger (Ch. 7)**	**Jarry (Ch. VI)**	**Drecker**	**Green (Gunther)**	**Severus Sebokht**[a] **Nau (Ch. IX)**	**Margoliouth (Gunther)**[b]
Cardinals	die vier Kardinal-punkte (p. 29, l. 6.11)	les quatre centres (p. 21, l. 1.7)	die vier Spitzen (p. 31, l. 19.23)	that / the four Central Points (p. 70, l. 13.16)	les ascensions (p. 284, l. 2)	the ascensions (p. 93, l. 12)
			die 4 Haupt-punkte des Astrolabs / *anguli* / *cardines* (p. 31, n. 13)	angles (p. 70, n. 1)	ces quatre centres / ces quatre lieux / pivots (p. 284, n. 3)	these four so-called centres / the four places / pivots (p. 93, n. 3)

Table (*cont.*)

Modern designation	Ioannes Philoponus Stückelberger (Ch. 7)	Jarry (Ch. VI)	Drecker	Green (Gunther)	Severus Sebokht[a] Nau (Ch. IX)	Margoliouth (Gunther)[b]
Ascendant	der Horoskop-Punkt (p. 29, l. 6–7.12–13.22)	le centre du levant (p. 21, l. 2.17) / (le) levant (p. 21, l. 7.14.22)	der Horoskopos (p. 31, l. 19.24, p. 32, l. 2)	the Ascendant (p. 70, l. 13.16) / the ascendant (p. 70, l. 23)	(le) centre de vie (p. 284, l. 3; p. 285, l. 14–15)	(the centre) of life (p. 93, l. 13; p. 94, l. 6)[c]
	der Aufgangs-punkt (p. 29, l. 26)		das Aufgehende (p. 32, l. 6)	the rising one (p. 70, l. 26–27)	les degrés de vie (p. 284, l. 6)	the degrees of life (p. 93, l. 16)
					(le) degré «de vie» (p. 286, l. 10)	the degree 'of life' (p. 94, l. 23–24)
					les ascensions (p. 285, l. 11.20.21–22.24, p. 286, 2.7.9–10)	the ascensions (p. 94, l. 3.14.17.21.23)
Descendant	der Unter-gangspunkt (West) (p. 29, l. 14.25–26b)	le couchant (p. 21, l. 9.20.22)	der Untergang (p. 31, l. 25) / das Unter-gehende (p. 32, l. 7)	the setting one (p. 70, l. 17–18.27)	les « descen-tes» (p. 285, l. 4; p. 286, l. 12.22.24; p. 287, l. 1)	the descents (p. 93, l. 14; p. 94, l. 25.32.34; p. 95, l. 1–2)
		le centre du couch-ant (p. 21, l. 20–21)	die Unter-gangsspitze (p. 32, l. 5–6)	the set-ting central point (p. 70, l. 25–26)	(le) degré «des noces» (p. 286, l. 12–13)	the centre 'of nuptials' (p. 93, l. 14–15)[d]
						the degree of conjunction (p. 94, l. 26)
Upper mid-heaven	der Mittags-punkt (p. 29, l. 7)	le milieu du ciel (p. 21, l. 26–p. 22, l. 1)	die Mitte des Himmels (p. 31, l. 19–20.24)	the Culminating (p. 70, l. 13.17)	(le) milieu du ciel (p. 284, l. 4; p. 285, l. 1.16.19.30, p. 286, l. 3.6.8.10)	the middle of the heaven (p. 93, l. 13–14.16–17; p. 94, l. 7.8.16.18.20–21.24)

Table (*cont.*)

Modern designation	Ioannes Philoponus Stückelberger (Ch. 7)	Jarry (Ch. VI)	Drecker	Green (Gunther)	Severus Sebokht[a] Nau (Ch. IX)	Margoliouth (Gunther)[b]
	Himmelsmitte (p. 29, l. 30)		die Spitze der Himmelsmitte (p. 32, l. 11)	the culminating central point (p. 70, l. 30)	(le méridien) (p. 286,1)	(the meridian) (p. 94, l. 17)
Lower mid-heaven	(die) Himmelsmitte unter der Erde (Nord) (p. 29, l. 14–15.30–p. 30, l. 1)	le milieu du ciel sous la terre (p. 21, l. 9–10, p. 22, l. 2)	die Himmelsmitte unter der Erde (p. 31, l. 26)	that beneath the earth to the culminating one (p. 70, l. 18)	celui qui est sous la terre (p. 284, l. 5)	that which is beneath the earth (p. 93, l. 15)
					(le) milieu de la terre (sur la moitié inférieure du méridien) (p. 286, l. 13–14; p. 287, l. 2)	the middle of the earth (on the lower half of the meridian) (p. 94, l. 26–27.34.37; p. 95, l. 2)
					(le) centre "des pères" (p. 286, l. 25)	the centre "of the fathers", i.e. the earth (p. 94, l. 35)[e]
					le centre du milieu de la terre (p. 286, l. 28)	

a In chapters 10 and 11, "l'horoscope" (p. 287, l. 6) / "the horoscope" (p. 95, l. 4.6), and "(le) degré de vie" (p. 288, l. 4; p. 289, l. 22–23) / "the degree of life" (p. 95, l. 21; p. 96, l. 19.25) are used for the horoscope; "(le) degré «des noces»" (p. 288, l. 4; p. 289, l. 23) / "(the degree of) nuptials" (p. 95, l. 21; p. 96, l. 19–20.25) are used for the descendant.

b According to McMahon, op. cit. (n. 4), p. 1045a, "an English version [...] from Nau's French".

c Compare "degree 'of life'" (p. 88, l. 28) and "'at the life'" (p. 88, l. 31; p. 95, l. 21; p. 96, l. 19.25).

d Compare "(degree) 'of nuptials'" (p. 88, l. 28–29) and "'at the nuptials'" (p. 88, l. 32).

e Compare "'at the fathers'" (p. 89, l. 3).

Appendix 3: Selection of Astrolabes with Planetary Symbols

Abbreviations:
LoT: lords of the triplicities
L: letters
S: symbols
W: words

Location:
AP: Adler Planetarium, Chicago
DAI: Dār al-Āthār al-Islāmiyya, Cairo
DB: Dār al-Baṭḥā, Fez
DS: Davids Samling, Copenhagen
MET: Metropolitan Museum of Art, New York
MHSO: Museum of the History of Science, Oxford
MIA: Museum of Islamic Art, Doha

References

Gunther: Robert T. Gunther, *The Astrolabes of the World*, Oxford, Oxford University Press, 1932, 2 vols.: v. 1.
King: David A. King, *In Synchrony with the Heavens: Studies in Astronomical Timekeeping and Instrumentation in Medieval Islamic Civilization*, Leiden, Brill, 2004–2005, 2 vols.: v. 2.

Maker	Place	Date	Terms	Faces	LoT	Location	Illustration
Ḥāmid b. ʿAlī al-Wāsiṭī	Baghdād (?)	9th / 10th c.	S	L	—	DAI 15352***	King 502, 8.2b
al-Khujandī	Rayy (?)	984–985	S	L	L	MIA SI.5.1999	King 507, 9b
Aḥmad and Muḥammad, sons of Ibrāhīm al-Iṣfahānī	Iṣfahān (?)	*c.*1100 (?)	S*	—	S*	MHSO, inv. no. 33767	King 521, 10c; Gunther XXII, XXIII
Badr, *mawlā* of Hibat Allāh al-Aṣṭurlābī	Iraq	1130–1131	S	—	L	AP A-84	King 54, 4.7.3

Table (*cont.*)

Maker	Place	Date	Terms	Faces	LoT	Location	Illustration
Ḥāmid b. Maḥmūd al-Iṣfahānī	Iṣfahān (?)	1152–1153	S, L	L	—	MIA MW.382.2007	King 49, 4.5.2
al-Ashraf ʿUmar	Yemen	1290	S	S	S	MET 91.1.535	King 626, 2.2
Muḥammad b. Jaʿfar b. ʿUmar, also called Jalāl	Kirmān (?)	1393–1394	W	S	L	DB 764***	King 773, 4b
Muḥammad b. Jaʿfar b. ʿUmar, also called Jalāl	Samarqand (?)	1426–1427	W	S, W**	W	D 25/1986	King 763, 1.6a

* The astrological tables are later additions; see King, 521–523.
** The planetary symbols do not correspond to the names of the planets given in the next line.
*** According to King.

CHAPTER 2

What's on the Back of an Astrolabe? Astrolabes as Supports for Planetary Calculators

Seb Falk

The life of Jean Fusoris (*c.* 1365–*c.* 1436) epitomizes the complex status and functions of astronomical instruments in medieval Europe. He was a prolific and highly influential craftsman, whose workshop in Paris produced astrolabes for kings and even the Pope.[1] As well as making instruments, he wrote treatises on their construction and use, revealing his constant drive to improve his designs. A canon of Notre Dame and master of medicine, tried for treason after he visited England during the Hundred Years War to collect an outstanding debt owed by the bishop of Norwich for an innovative equatorium, he personifies the intersections between the sciences of the stars, the body and religion; the court, the university and the church.

At his trial, Fusoris revealed that he had discussed "the subject of astrology" with Henry V "by means of the bishop of Norwich", and had then given the English king an astrolabe, together with a manual for its use.[2] However, he also claimed that, when shown a chart for the king's nativity by "an English master of theology", he had refused to interpret it, stating that "he did not involve himself in judgements, but was certainly interested in motions".[3] The distinction Fusoris was making was well established: on one side was legitimate astrological study of celestial configurations, possibly using instruments like astrolabes; on the other were more reprehensible efforts to make predictions based on those configurations at the time of the subject's birth. This distinction had been stated clearly by Thomas Aquinas (1225–1274), who admitted that "celestial bodies have an effect on bodies ... [and] on powers of the soul that are acts of bodily organs", but admonished:

1 Emmanuel Poulle, *Un constructeur d'instruments astronomiques au XVe siècle: Jean Fusoris*, Paris, Librairie Honoré Champion, 1963, pp. 2–5.

2 Procès de Me. Jean Fusoris, 1415–1416: Archives Nationales, Paris, LL//85, quoted in Léon Mirot, "Le Procès de Maître Jean Fusoris", *Mémoires de la société de l'histoire de Paris et de l'Ile-de-France* 27 (1900), pp. 137–287: 174.

3 *Ibid.*

 | DOI:10.1163/9789004381438_004

> if anyone were to use observation of the stars to foreknow fortuitous or chance future events, or to know future human actions with certainty, his actions are based on a false and empty belief; and so the work of the demon creeps in, whence it will be a superstitious and unlawful divination. On the other hand, if anyone were to use observation of the stars to foreknow future things that are caused by heavenly bodies, such as drought or rain or others of this sort, it will be neither an unlawful nor a superstitious divination.[4]

Astrology experienced a rise in respectability in the thirteenth and fourteenth centuries, to the point where deterministic prediction seemed quite acceptable, especially in courtly settings.[5] By the later fourteenth century, according to one historian, "the theology of astrology was in disarray".[6] Nonetheless, as Fusoris's account indicates, a distinction between licit and illicit astrology was still widely recognised.

Fusoris's testimony also illustrates the social importance of instruments such as astrolabes. Although he is now best known as an astrolabe-maker—29 extant astrolabes were either made by him or directly influenced by his designs—his compulsion to innovate caused him both to critique established practices in astrolabe production and to make other instruments which better fitted his interests.[7] This chapter will begin by assessing some of Fusoris's critiques, and will proceed to examine two late medieval instruments, the first perhaps designed by Fusoris, which present alternative uses for the back of an astrolabe. Each instrument provided a tool for finding planetary positions, which, besides being of interest to astronomers like Fusoris, were essential for casting the kind of nativity he disavowed. One of these instruments is extant in brass, while the other exists only as a textual description. The discussion will address the relationships between texts and objects, as well as the way in

4 Thomas Aquinas, *Summa Theologiae*, 1.115.4; 2.2.95.5, ed. Fundación Tomás de Aquino, www.corpusthomisticum.org/iopera.html (accessed 30 November 2016); Aquinas was drawing on St Augustine, *De civitate Dei*, v. 6. For background on the status of astrology in medieval theology, see Theodore Otto Wedel, *The Mediaeval Attitude toward Astrology, Particularly in England*, New Haven, Yale University Press, 1920, pp. 63–89.

5 John North, "Astronomy and Astrology", in D.C. Lindberg, M.H. Shank (eds.), *The Cambridge History of Science, Vol. II: Medieval Science*, Cambridge, Cambridge University Press, 2013, pp. 456–484: 474.

6 Hilary Carey, *Courting Disaster: Astrology at the English Court and University in the Later Middle Ages*, Basingstoke, Macmillan, 1992, p. 15.

7 David A. King, "An Ordered List of European Astrolabes to *ca.* 1500", in D.A. King, *Astrolabes from Medieval Europe*, Farnham, Ashgate, 2011, XII, pp. 8–10.

which theoretical designs were realized within the practical constraints imposed by size and material.

Planetary instruments could exist in a variety of formats, as individual devices or as parts of multifunctional compendia. This chapter will consider the basis for the choices made by their creators. Just as an important feature of manuscript studies is consideration of the materials used by the scribe—parchment or paper, iron or oak gall—so, it will be argued, we should give more attention to the *supports* used for instruments: the material forms in which they were necessarily produced and which may have had an effect on their use, but which were not essential to their theoretical functions. The question of supports for equatoria and other planetary instruments is often ignored, perhaps because so many survive only as designs or descriptions. The fact that they were often little more than physical diagrams makes it easy to forget that they are three-dimensional.[8] I will show that the choice of support entailed inevitable compromise, but also enabled considerable innovation.

It will also be important to question the notion of a "complete" or "finished" instrument. It is widely accepted that objects are not immutable, and that we can learn a great deal from the changes made after they are first produced. But notions such as the "wounded artefact", while helpful in reminding us that objects can tell us much when they are damaged, and were not always used in the way their creator intended, may also create the impression that there is necessarily a perfect, "healthy" artefact.[9] This obscures the fact that astronomical instruments were frequently altered or customized by their owners. Rather than assuming the possibility of a standard "complete" astrolabe, we should be careful to examine objects in their specific contexts of production, modification and use.[10]

New Uses for Old Astrolabes

Fusoris wrote a treatise on the astrolabe, which provides a fascinating overview of the craftsmanship employed at what was probably the first workshop to undertake anything like large-scale production of those instruments.[11] His most detailed discussion of the back of the astrolabe appears in a treatise

8 John D. North, *Richard of Wallingford: An Edition of His Writings*, Oxford, Clarendon Press, 1976, 3 vols.; v. 2, p. 261.

9 Stephen Greenblatt, "Resonance and Wonder", *Bulletin of the American Academy of Arts and Sciences* 43 (1990), pp. 11–34: 22.

10 Bill Brown, "Thing Theory", *Critical Inquiry* 28 (2001), pp. 1–22.

11 Jean Fusoris, "Composition de l'astrolabe", Text 1 in Poulle, op. cit. (n. 1), pp. 95–108.

on his equatorium. Written to accompany the instrument he sold to Richard Courtenay, Bishop of Norwich, it reveals much about the ways in which Fusoris viewed the two instruments in relation to one another. A particularly severe critique of the astrolabe's standard calendar scales appears when he discusses methods of modelling the Sun's motion:

> [...] the instrument of the Sun can be made in many different ways. The first is the crude method by which it is commonly made on the back of an astrolabe; but it is true that this method suffers from many shortcomings. The first is because the deferent of the Sun and its centre are not moved by the motion of the eighth sphere, but always remain aligned with the same part of the zodiac. The second and larger shortcoming is because that method assumes that the Sun on its deferent traces out the whole zodiac in precisely 365 days, which is not the case.[12]

He goes on to explain that the usual astrolabist's approach of making the instrument for the second year following a leap year is an unsatisfactory approximation, and proposes the addition of marks to the alidade to allow the user to adjust his reading of the calendar for different years in the leap cycle. However, even this measure is unsatisfactory for Fusoris, since, he notes, the addition of one intercalary day every four years is excessive (by 1' 46" every four years). To compensate, he proposes that the fiducial line of the alidade be filed down, "just as the zodiac of an astrolabe rete is commonly filed down".[13] He then goes on to describe a preferable solution, in which the Sun's eccentric is carried on a mobile disc, but here he tacitly reverts to discussing his equatorium rather than astrolabes. This discussion reveals not only the ways in which the boundaries between instruments were blurred in the mind of the craftsman, but also how he saw astrolabes not as fixed or finished, but subject to adjustment and recalibration. Fusoris's remark about the common practice of filing down the rete shows that he was not alone in this, and indicates that he, like others, was aware of the inevitable limitations of some of the traditional features of the astrolabe, leaving the way open to experimentation with alternative uses of the back.

Fusoris's critique lends credence to the suggestion that an instrument providing alternative employment for the back of an existing astrolabe may also

---

12 Jean Fusoris, "Liber primus de motibus planetarum per instrumenta manualiter mota", Text 3 in Poulle, *ibid.*, pp. 125–180: 150. See the comparisons made by Richard of Wallingford between his Albion instrument and the astrolabe; North, op. cit. (n. 8), v. 1, p. 331.

13 Fusoris, "Liber primus de motibus planetarum", cit. (n. 12), p. 152.

be his work. The instrument is described in a treatise with incipit "Motus medii planetarum", which was thought to survive in four copies and has never been satisfactorily dated or attributed.[14] However, a fifth copy (perhaps unnoticed until now simply because the first two words are transposed) may provide a link with Fusoris. It appears immediately after a table listing stars for use on an astrolabe; the table is attributed to Fusoris and dated 1428.[15] The short, but creative and clear treatise explains the construction and use of a handy dial to be fixed "on the back of an astrolabe".[16] Briefly put, the dial contained a series of concentric circles.[17] Each space between two concentric circles was graduated for the annual or daily mean motion or mean argument of a planet (including the Sun and Moon) or *Caput Draconis* (the lunar ascending node); the graduations all started in the same place but, as each element takes a different time to complete one revolution, each circular graduation was equivalent to a different duration. The whole device was intended to rotate freely on the back of an astrolabe, so that it could be zeroed at the appropriate place on the astrolabe's degree scale (which thus functioned as an ecliptic scale). It was read by stretching a thread over the mark for the desired time on each circle, and reading off the corresponding motion on the scale. All in all, the device was an intriguing fusion of tables and instruments. It contained data from tables, arranged spatially in a way that made it akin to a diagrammatic planisphere of the Ptolemaic heavens. It was a substitute for tables, yet had something of the durability of an instrument, with its layout representing the eternal circularity of planetary motions.[18] And although it was essentially only strips of data cells arranged around a central point, still perhaps considered a table by the author of the treatise (his word "tabula" is delightfully ambiguous, able to signify any or all of a board, flat surface or arrangement of data), it also edged from

14 Leipzig, Universitätsbibliothek, Cod. 1469, fols. 237r–240r; Munich, Bayerische Staatsbibliothek, Clm 19689, fols. 162v–164v; Oxford, Corpus Christi College, 152, fols. 276v–279r; Wolfenbüttel, Herzog August Bibliothek, 2816, fols. 140v–141v.

15 Universidad de Salamanca, Ms. 2621, fols. 10r–11v. Guy Beaujouan, *Manuscrits scientifiques médiévaux de l'université de Salamanque et de ses "colegios mayores"*, Bordeaux, Féret, 1962, p. 60. An alternative candidate is John Simonis of Selandia, whose equatorium treatise "Speculum planetarum", written at Vienne in 1417, appears in the same codex as four of the five copies of "Medii motus planetarum".

16 Ms. 2621, cit. (n. 15), fol. 10v.

17 For a detailed explanation of its workings, see E. Poulle, *Les instruments de la théorie des planètes selon Ptolémée: équatoires et horlogerie planétaire du XIIIe au XVIe siècle*, Geneva, Droz, 1980, pp. 106–110.

18 If, as the author of the treatise suggests, the radices were engraved onto spare space on the back of the astrolabe, the need for separate tables could be eliminated. Ms. 2621, cit. (n. 15), fol. 11r.

astronomy into cosmology: the circles were arranged from Saturn on the outside to the Moon within, and inside the circle of daily mean lunar motion the author placed the spheres of fire, air, water and earth.

Even if the author considered his creation to be a table, it was not divorced from the astrolabe that functioned as its support. It not only made use of the astrolabe's 360° scale for its construction and use, but was also laid out with reference to an astrolabe, quartered by diameters whose ends were described as "meridional", "midnight", "western" and "eastern". These terms were meaningless in the context of the mean motion dial, but the author apparently found them useful in thinking about its design. In the same way, he introduces the possible scale of the instrument in the first sentence of the treatise by stating that "it may also be the same size as a *tabula* with almucantars", thus drawing an analogy between the circles on an astrolabe plate and those on his device.[19] More generally, the author highlights his conformity with the conventions of an established genre of instrument treatises, by remarking, for example, that ABCD, which designate the diameter ends, are "letters often used". He expects his reader to be equally familiar with the genre: his injunction that "the circle of signs is not required on this *tabula*, since it is unnecessary for its use", is clearly aimed at fellow makers who might automatically divide the rim of a circle into 360 degrees because they had not realized that co-opting the astrolabe's scale made it unnecessary.[20]

The author caters for the abilities and preferences of these craftsmen-readers, instructing them, for example, to "make a widely spaced circle, in which you can write the names of months and of whichever feast days you wish".[21] This glimpse of choices in the design process is rare in treatises that usually present instruments as finished. The choice of which religious feast days to include in the calendar was one way in which the makers of otherwise conventional astrolabes could personalize their creations; a number of surviving instruments feature the days of unusual, regionally specific saints alongside those that appeared in standard calendars.[22] Still greater flexibility is offered by the final sentence that appears in three of the five copies of the treatise.[23] Here the author envisages an alternative instrument with two "figures" which

19 *Ibid.*, fol. 10v.

20 *Ibid.*

21 *Ibid.*

22 John Davis, Michael Lowne, "An Early English Astrolabe at Gonville & Caius College, Cambridge, and Walter of Elveden's Kalendarium", *Journal for the History of Astronomy* 46 (2015), pp. 257–290: 272.

23 Corpus Christi College, 152, cit. (n. 14), fol. 279r; Wolfenbüttel, 2816, cit. (n. 14), fol. 141v; Salamanca, 2621, cit. (n. 15), fol. 11v.

"can be made on one *tabula*". The first would be the planetary device described in the treatise; the second an equivalent instrument to give the locations and relative positions of places on Earth. The astrolabe has disappeared, yet its layout is referenced in the way the two instruments are summarized: "according to convention, since just as one [side] measures the heavens, the other measures the Earth".[24] In exactly the same way, the front of an astrolabe with its net of stars looked towards the heavens, while the back, with its calendar of feast days and surveyor's shadow square, represented terrestrial things. The interrelationship of different instruments was such that mental vestiges seem to persist, even after all physical traces were removed.

The usefulness and flexibility of the device described in the "Medii motus planetarum" treatise must be what led the treatise to be copied, and its instructions followed, until at least the sixteenth century.[25] Petrus Apianus was later to incorporate a calendar of mean motions into his lavish *Astronomicum Caesareum* (1540); Fusoris himself had already included such a device in the design of his equatorium compendium. Nevertheless, it should be stressed that the "Medii motus planetarum" treatise does not yet describe an equatorium. It did not compute the positions of the planets, but rather was a handy source of the basic data necessary for such computation. Some of this data is already simplified: the annual motion of the Sun's mean argument is omitted, which implies that the Sun starts from the same point each year.[26] Such simplification saved time and material, but would not have satisfied all astronomers. Those seeking a more complex, comprehensive alternative might have preferred to see a complete equatorium as an integral part of the back of an astrolabe. It is to a rare example of this that we now turn.

An Oxford Astrolabe-Equatorium

Merton College, Oxford, is fortunate to possess two fourteenth-century astrolabes. The later of these (*c.* 1390) is particularly noteworthy for the way it has

24 *Ibid.*

25 Corpus Christi College, 152, cit. (n. 14), fols. 276v–279r, copied by Nicolaus Krazer, *c.* 1523. On Kratzer, see J.D. North, "Nicolaus Kratzer, the King's Astronomer", in North, *Stars, Minds and Fate*, London, Hambledon Press, 1989, pp. 373–400: 388–389. A device operating on similar lines survives on the back of a sinecal quadrant (made *c.* 1500) at the Museum of the History of Science, Oxford, inv. no. 16856.

26 This might be taken as evidence against Fusoris's authorship, since he discussed the problems of the leap cycle. But his output was varied; his critique of astrolabe calendars did not stop him producing dozens of them.

been modified by the addition of a lunar dial on the back. The earlier, which may be the "larger astrolabe" left to Merton by Simon Bredon in 1372, has long interested historians.[27] Its technical features have been discussed in some detail by Emmanuel Poulle so, save for taking up some of the points Poulle raises, we shall not delve too deeply into aspects of the design of the equatorium.[28] What is pertinent here is the way in which the technical design has been materialized within the constraints of the back of an astrolabe.

The most noticeable aspect of the front of the Merton astrolabe-equatorium, aside perhaps from the fact that it is for a single latitude (having no separable plates or womb to hold them), is that the maker has placed the usual contents of an astrolabe back on the limb of the front (*Fig. 2.1*). Working from the outside inwards, they consist of a scale numbered both for measuring altitudes and shadow-square computation (this was also marked by a later user with some hours in Roman numerals); a concentric calendar of months and days; a scale of zodiac signs and degrees; four circles for use in different years of the leap cycle (a note on the outer limb explains their usage); and an inner scale of 360° with hour markings every 15°. The outer scale contains 720 divisions but, since angles are measured via the plumb-line at the limb, they are halved and the scale thus reads to a quarter-degree; the numbering reflects this. The use of a plumb-line, together with the sighting pinnules that protrude from the limb, mirrors the arrangement on the Albion designed by Richard of Wallingford in 1326.[29] It is thus not surprising to find that the Albion is one of four instruments cited in a treatise, written in Oxford around 1349, that appears to describe the Merton equatorium, or an instrument much like it.[30] After a historical introduction taken with some adaptations from the "Equatorium" of John of Linières (the two treatises share the incipit "Quia nobilissima scientia astronomie"), it is divided into a first, ostensibly theoretical, section, followed

27 Robert Taylor, "Remarks on an Astrolabe, Belonging to F.A. Hyett, Esq., of Painswick House", *Transactions of the Bristol and Gloucestershire Archaeological Society* 12 (1887–1888), pp. 6–23. Bredon's will is printed in Frederick M. Powicke, *The Medieval Books of Merton College*, Oxford, Clarendon Press, 1931, pp. 82–86. For the astrolabe's exhibition at the Royal Society by Derek J. de Solla Price, see Seb Falk, "The Scholar as Craftsman: Derek de Solla Price and the Reconstruction of a Medieval Instrument", *Notes and Records of the Royal Society* 68 (2014), pp. 111–134.

28 Poulle, op. cit. (n. 17), pp. 200–205.

29 North, op. cit. (n. 8), v. 1, p. 299.

30 Cambridge University Library, Gg.VI.3, fols. 217v–220v (*c.* 1349); Oxford, Bodleian Library, MS Digby 57, fols. 130r–132v (*c.* 1375), edited and translated in Seb Falk, "A Merton College Equatorium: Text, Translation, Commentary", *SCIAMVS* 17 (2016), pp. 121–159. The other citations are to the instruments of Campanus of Novara (*c.* 1220–1296), John of Linières (*fl. c.* 1320) and Profatius Judaeus (*c.* 1236–*c.* 1304).

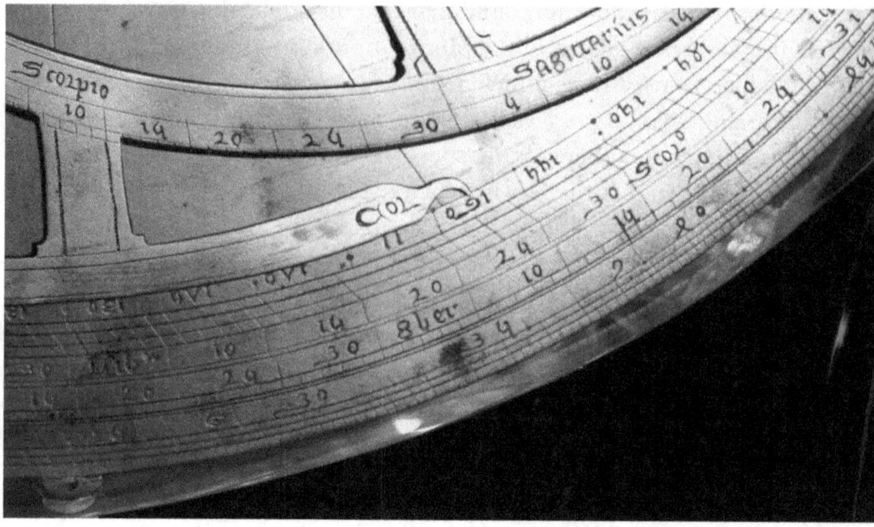

FIGURE 2.1 Astrolabe-equatorium, *c.* 1350; detail of limb (Merton College, Oxford, SC/OB/AST/2; reproduced by permission of the Warden and Fellows of Merton College, Oxford).

by a more practical part discussing the uses of the equatorium.[31] However, this dichotomy is misleading and overlooks the striking symbiosis between theory and practice represented by treatise and instrument.[32] The theory is instrumental in both senses of the word: it is the minimum necessary theory for successful operation of the instrument, directed to that purpose. We see this in the style of explanation; for example, the mean centre of any planet is defined as "the distance between the apogee of the equant and *a thread* stretched from the equant centre to the epicycle centre".[33] The author appears quite conscious of this striking blurring of boundaries between Ptolemaic theoretical model and physical instrument, and seems to understand how both work towards increased understanding. In the final sentence of his introduction, he states that "through use of the instrument of Campanus, Linières or [Profatius] Judaeus, the *theorica* may be put forth so that its effect may be sufficiently plain".[34] The Latin word *theorica* refers to a geometrical model, something already very close to the equatoria that were devised as three-dimensional representations

31 The shared incipit is probably the reason why the Gg.VI.3 scribe ascribed the text to John of Linières.

32 The structure of the text is discussed in North, op. cit. (n. 8), v. 2, p. 272; Poulle, op. cit. (n. 17), pp. 200–201; and Falk, op. cit. (n. 30).

33 Gg.VI.3, cit. (n. 30), fol. 218r; my emphasis.

34 *Ibid.*, fol. 217v.

of it. So for the author, the use of the instrument was a vital part of expounding the astronomical theory. Whether "effect" refers to the instrument's functions or the theory's effects becoming clear is delightfully, perhaps deliberately, ambiguous in that passage. The whole treatise is quite revealing of the role that instruments played in elucidating theories; the very physical description of planetary stations and retrogradations must surely have been written by someone who had an equatorium in front of him. Of course, instruments could perform this function while also providing numerical results for practical purposes.

The author's use of the singular "instrument" indicates his desire to stress the similarity of the three instruments he cites; he later writes that "there is little difference in the method of using them".[35] However, the three differ substantially, from each other and from his description.[36] If our author is describing the Merton equatorium—or any other single instrument—perhaps this is "the other, newly composed, equatorium", which he mentions in his conclusion.[37] Perhaps he did not introduce it earlier because it could not be attributed to a famous inventor, something that was clearly a concern for this and other writers. His description is close to the Merton equatorium not only in design, but also in time and space. A ready reckoner for precession engraved on the back suggests that the equatorium was made in 1350; on the front we find "Lat' 52. 6m. Oxonia".[38] The two copies of the treatise mention radices and tables computed for 1348 and 1350 at Oxford.[39] It explains the use of these radices and tables, with appropriate techniques for adjustment to different longitudes and interpolation between tabulated values, as a crucial preliminary step before the instrument itself was brought into play.

The first step in its use was to place it on a large, flat table. Here we encounter a further advantage of the use of sighting pinnules on the front. The equatorium clearly had to be horizontal in order for the threads to be carefully moved into position to give the longitudes of the planets, but this meant that when the instrument-maker transferred the features of the back onto the front, he

35 *Ibid.*, fol. 220v.

36 Francis S. Benjamin, G.J. Toomer (eds.), *Campanus of Novara and Medieval Planetary Theory*, Madison, University of Wisconsin Press, 1971, pp. 141–143; "Composito equatorii secundum Johannem de Lineriis", in Derek J. Price, *The Equatorie of the Planetis*, Cambridge, Cambridge University Press, 1955, 188–196; Fritz S. Pedersen (ed.), *Petri Philomenae de Dacia et Petri de S. Audomaro opera quadrivialia*, Copenhagen, Societas Linguae et Litterarum Danicarum, 1983, p. 43; Poulle, op. cit. (n. 17), pp. 205–210.

37 Gg.VI.3, cit. (n. 30), fol. 220v.

38 The more generally accepted latitude of Oxford was 51° 50'.

39 The earlier copy states that the longitude of Oxford is 18°; the later gives 15°, with a latitude of 51° 56'.

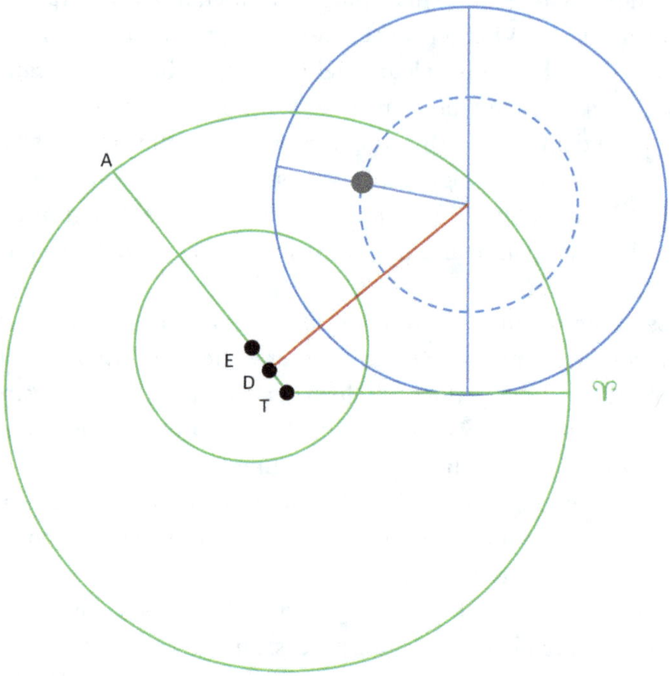

FIGURE 2.2 Diagram of the Merton astrolabe-equatorium, showing the probable workings of the epicycle and common deferent radius.

could not incorporate an alidade, as the sights would have prevented its lying flat. The use of pinnules was a neat solution to this problem.

The treatise explains briefly how to use the instrument to compute the longitudes of Saturn, Jupiter, Mars and Venus, before going into greater depth on the more complex procedures for Mercury and the Moon. The instrument included a common epicycle for use with all the planets. This part does not survive, but its workings can be reconstructed from the treatise, and numerous scratches on the face of the instrument are testament to its use. The epicycle was connected to the face of the instrument by a common deferent radius (*Fig. 2.2*). One end of this was fixed to the centre of the common epicycle, while the other end was to be pinned to the deferent centre of the desired planet. In the case of Mercury, whose deferent centre moved on a small circle, the treatise describes a circle of 36 small holes, in one of which the pin was placed; this circle survives on the Merton instrument (*Fig. 2.3*). The common epicycle almost certainly took the form of a ring rather than a disc, so as not to obscure too much of the face of the instrument (not to mention saving materials). The

FIGURE 2.3 Astrolabe-equatorium, *c.* 1350; back (Merton College, Oxford, SC/OB/AST/2; reproduced by permission of the Warden and Fellows of Merton College, Oxford).

part on which the planets' epicycle radii must have been marked is described simply as the "true epicycle". This might lead one to suppose that the common epicycle could have been a disc with concentric circles for the planets' epicycles, but the way the treatise describes the use of this part makes it more probable that it took the form of a rule. If so, "true epicycle" is a description of the object's function, rather than its form. Here, too, the boundaries are blurred between description of the constituent parts of the instrument and their roles as symbols in planetary models.

As well as the common epicycle, the Merton equatorium is also missing a disc for the more complex model of the Moon. The extant equatorium has holes that might have been used to attach such a disc, and the treatise describes its use. The phrasing of this description, with statements such as "the plate is fixed to the rule by its wedge", strongly suggests that the author was describing a particular physical object. This undermines his claim to be writing canons for various different instruments; perhaps we should assume he added that introductory claim to widen the usefulness of his treatise, without really modifying its main content.

In order to locate the epicycle centre on the ecliptic, the Merton design uses the mean centre for each planet. Because these are measured from the planets' apogees, each planet needed a separate graduated equant circle.[40] These circles created problems for both maker and user. For the maker, graduating eight eccentric circles (for the ecliptic, the eccentric of the Sun, the equants of Mercury, Venus, Mars, Jupiter and Saturn, and the little circle of Mercury) would have been a laborious and error-prone process. For the user of these circles, their visibility would have been reduced when the common epicycle was laid over the face, making the operation of the threads somewhat tricky. Nevertheless, the maker apparently took steps to make the equant circles as readable as possible. First, because the radii are irrelevant to the model, he was able to nest them, making the best possible use of the space on the face of the instrument. Secondly, in order to minimize confusion in the use of the circles and closely clustered deferent centres (where the common deferent radius had to be attached), he labelled each circle with the name of the planet at its apogee. This allowed the apogee, the starting point for laying out the mean centre, to be found quickly. He then ingeniously engraved the labels marking each deferent centre the same way up as the name of the corresponding planet, thus eliminating the possibility of using the wrong deferent centre. Finally, despite their varied sizes, the equant circles are all graduated in 360 degrees and numbered in fives in this way: 5, 10, 15, 20, 25, 1, 5, 10, 15, 20, 25, 2, etc. This allows numbers of signs to be found easily.

One curious feature of the equant circles is that the instrument has separate circles for the Sun and Venus, with separate holes for the deferent centre of the former and equant centre of the latter, though most readings of the Ptolemaic theory would make them identical. This could be considered a weakness of the design, but it is quite understandable if we consider that the position of the Sun could be found without the common epicycle (as the treatise explains),

40 An alternative approach is to use the mean longitude, measured from the vernal point. This allowed instrument-makers to avoid engraving equant circles, but did require users to translate longitudes measured at the Earth to the appropriate equant centre using parallel threads, which was a potential source of error.

so that its eccentricity need not be scaled to conform to the common deferent radius.[41] If the maker wished to have separate circles to emphasize the separation of the Sun and Venus, he was free to do so; the consequent separation of their centres was surely a small price to pay.

The circles and apogees are, of course, oriented with respect to the zodiac. But the zodiac could be oriented in any way. The equatorium, it should be stressed, was a calculating device, not an observational one, so its orientation was unimportant. Since the corresponding part of the front of the astrolabe was the rete, which turns, there was no obviously symmetrical solution. The invariable practice on astrolabes with eccentric calendars (which, as we have seen, represent a solar equatorium) was to place the Sun's apogee at (or near) the top of the instrument, making an implicit connection between the apogee and the meridian. Since, according to the data in the Alfonsine Tables, the Sun's apogee was within 1° of the head of Cancer between 1278 and 1478, this would leave the vernal point (head of Aries), where most calculations began, almost exactly at the "west" point on the right hand side. Most equatorium designers followed this convention.[42] However, the Merton maker chose to place the vernal point at the top of the instrument. This divergence from the more common practice suggests two things about his priorities. First, the roughly 18' difference between the apogee and the head of Cancer in 1350 was not negligible to him. On the instrument we can clearly see that the apogee of the Sun and Venus, where the numbering of their equant circles begins, does not line up with the head of Cancer on the ecliptic scale (*Fig. 2.4*). Although the ecliptic was only graduated in degrees, a fraction amounting to almost a third of a degree was clearly considered significant by the maker. His awareness of precession is further attested by the presence of a small table engraved on the disc, showing the motion of the eighth sphere in increments of 20' at 33-year intervals from 1350 to 1450 (*Fig. 2.3*); the same value is given in the "Quia nobilissima scientia astronomie" treatise, which states that "the motion of the eighth sphere is almost exactly 1° in 100 years".[43]

Secondly, taking the focus away from the Sun broke the link between equatorium and astrolabe: the maker appears to have seen his creation as two quite separate instruments. He thus rejected any temptation to draw an analogy between the astrolabe's planispheric map of the heavens and the chart in the plane of the ecliptic represented by the equatorium. It seems that he

41 Poulle, op. cit. (n. 17), p. 202, considers this "heretical".

42 See, for example, astrolabe and equatorium, late-fifteenth century, Museum of the History of Science, Oxford, inv. no. 49847.

43 Gg.VI.3, cit. (n. 30), fol. 219v.

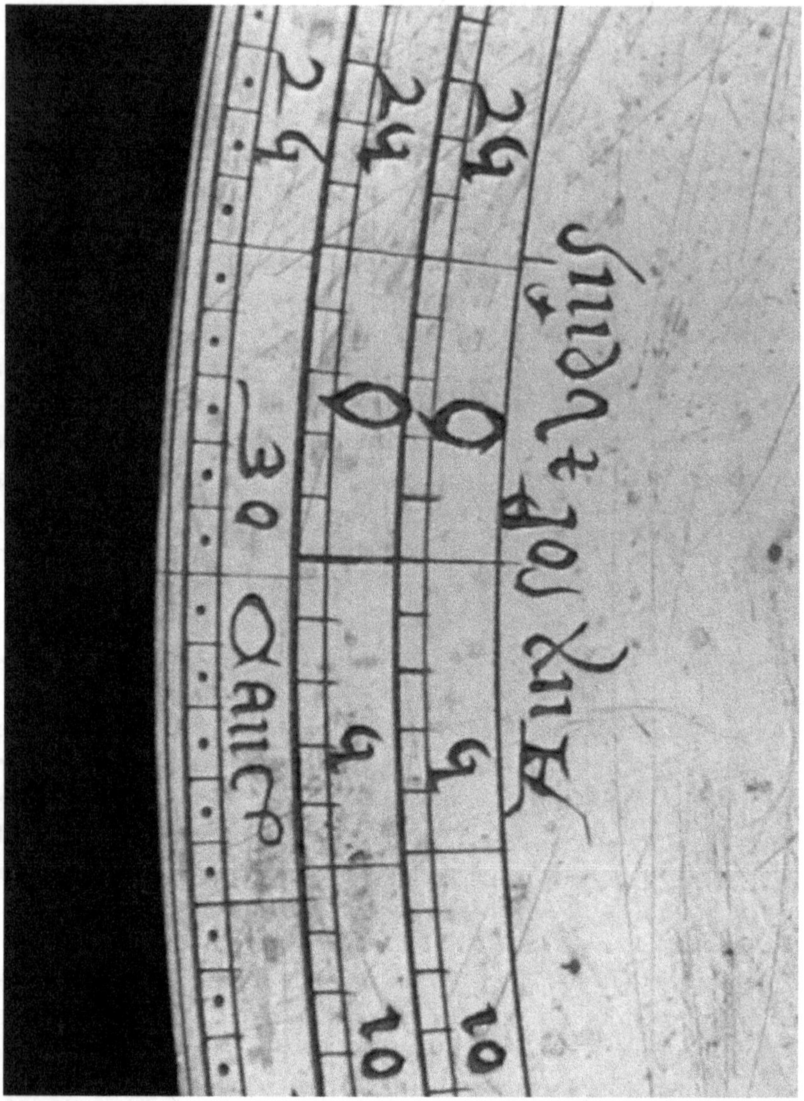

FIGURE 2.4 Astrolabe-equatorium, *c.* 1350; detail of the back (Merton College, Oxford, SC/OB/AST/2; reproduced by permission of the Warden and Fellows of Merton College, Oxford).

saw the equatorium not as a planetarium, but simply as a calculating device. By putting the vernal point at the top of his instrument, the Merton maker ensured that calculation would begin in the same place on both sides. This focus on numbering, rather than the Sun, allows the Merton equatorium to convey a numerical rather than spatial understanding of astronomy. This was an instrument that had little to do with stellar observation, and far more to do with computation using data from tables.

The Merton astrolabe-equatorium may have contained two quite separate instruments, but was nonetheless part of a cohesive, carefully planned whole. The transposition of the scales usually on the back of the instrument to the front, and the use of pinnules rather than an alidade with sights, freed the back to contain a clear, user-friendly calculating device. There were some drawbacks to such a compendium: apart from the limited visibility of the equant circles already mentioned, the holes at the deferent and equant centres, to which the threads and common deferent radius were attached, are pierced right through the instrument, though this is hardly noticeable and does not prevent measurement of altitude and azimuth on the astrolabe. The critical compromise was in size: at 362 mm in diameter, this instrument is as large as it can possibly be while keeping the astrolabe viable for observation; its small throne and the arrangement of the pinnules mean that two people were needed for easy sighting, but it was not uncommon for observation to be a collaborative activity. Thus, notwithstanding its incomplete state now, the Merton compendium and treatise are our best evidence for how the back of an astrolabe could be used as a support for an entirely brass equatorium in the medieval period.[44]

Conclusion

Instruments were born of such compromises. John of Linières had complained of the complexity and weight of Campanus of Novara's equatorium, which made it so difficult to make and move. The author of the "Quia nobilissima scientia astronomie" treatise, when copying John's introduction, added his own complaint about the difficulty, lengthiness and tedium of the necessary supporting calculation with tables.[45] Even Jean Fusoris, as we have seen, was willing to produce compromised instruments while still noting their deficiencies. Each of the instruments examined in this chapter responds in its own way to such complaints. Neither provides a complete solution: the Merton equatorium still required tables, whose preliminary use is explained at great length in "Quia nobilissima scientia astronomie", and the instrument described in

44 The design should be compared with the *Equatorie of the Planetis*; Price, op. cit. (n. 36); http://cudl.lib.cam.ac.uk/view/MS-PETERHOUSE-00075-00001 (accessed 6 April 2018). This was specifically 72 inches in diameter. See also the late-fifteenth-century astrolabe-equatorium, Museum of the History of Science, Oxford, inv. no. 49847.

45 "Compositio equatorii", cit. (n. 36), p. 188; Gg.VI.3, cit. (n. 30), fol. 217v.

"Medii motus planetarum" replaced tables but only provided mean longitudes, not true positions. Their focus and the nature of the compromises reveal much about the priorities of their creators. The primary concern may have been the balance of speed and precision required by users: while there was no limit to the precision a mathematically minded astronomer might pursue using tables, an astrologer compiling a horoscope needed only to work in degrees except in borderline cases, and thus would want to dispense with tables as far as possible in order to take advantage of the speed—and potential to impress his client—offered by an instrument. In more educational environments the priority may have been visual clarity, in order to represent, elucidate and perhaps teach astronomical and astrological theories. Whatever the balance of priorities, makers worked within the constraints of materials and space. They also bore in mind the practical concerns of sale: some instruments can be shown to have been made to purchasers' specifications; others suggest that innovation was supplier-driven, as makers sought competitive advantage; and still others were perhaps made in pursuit of innovation for its own sake, without regard to profit.[46] This last tendency can be seen across mathematical disciplines; for example, hand diagrams used in manuscripts to illustrate finger-calculating methods show constant innovation across the Middle Ages, as each scholar tried to invent his own method or improve existing ones. It could be argued that the variety in the surviving instruments from this period indicates the same drive towards innovation for its own sake. This innovation in instrument design was to continue, and indeed accelerate, through the first part of the early modern period.[47]

This chapter's analysis of the ways in which the backs of astrolabes were used as supports for planetary instruments, whether for astrological or other purposes, has wider implications for the way we view instruments. Those we have examined blur the boundaries between presentation and computation of data, between demonstration of theories and practical observation, between astronomy and cosmology. This should encourage us to reconsider the sometimes narrow categories into which we place objects; classes like "astrolabes" and "equatoria" are intended to help us understand the nature and purpose of texts and objects, but they can be unhelpful and their boundaries should

46 Davis and Lowne, op. cit. (n. 22), pp. 271–276. Fusoris's account indicates his conflicting motivations; see Mirot, op. cit. (n. 2), pp. 232–234.

47 An astrolabe-equatorium, Museum of the History of Science, Oxford, inv. no. 49847, from the dawn of this period exemplifies this; Anthony J. Turner, "From Brass to Text: The European Astrolabe in Literature and Print", in Koenraad Van Cleempoel, *Astrolabes at Greenwich*, Oxford, Oxford University Press, 2005, pp. 31–40.

be tested.[48] More generally, the identification of instruments that function as tables or guides to theory blurs the text-instrument boundary. A text can be an instrument; a *theorica* can sometimes seem as tactile as a physical model; a brass or wooden object might be little different from a diagram. This is important not so much for our definitions as for our understanding of how medieval astronomers saw texts and objects. For example, the question of why so few physical equatoria survive, in contrast to the many surviving treatises describing them, is complicated if blurring of categories makes that contrast invalid. The fact that Campanus of Novara's description of his equatorium was known simply as a *Theorica planetarum* highlights immediately the mixed meaning of the first word, and the fact that texts may not be all they seem: some that appear theoretical have a great deal to tell us about instruments, while others that describe instruments in very specific terms could have had a predominantly theoretical, pedagogical purpose.

Historians reading instrument texts inevitably ask whether they were intended to be made. If so, their authors must have considered the possible supports that could be used. This is not purely a craft question, but goes to the heart of the purpose of any planetary calculator. The practical benefits of saving time, effort and material by using the back of a portable astrolabe must have been balanced against the precision achievable with a larger, less portable instrument. Theoretical or pedagogical considerations, including the link between the two instruments, must have been considered. The same design could serve multiple purposes, but one or more of those purposes could be prioritized by the mode of presentation. Thinking through these complex relationships should bring us to a deeper understanding of what was in the minds—and hands—of medieval astronomers when they designed, described, made, modified and used instruments.

48 Adam Mosley, "Spheres and Texts on Spheres: The Book-Instrument Relationship and an Armillary Sphere in the Whipple Museum of the History of Science", in Liba Taub, Frances Willmoth (eds.), *The Whipple Museum of the History of Science: Instruments and Interpretations*, Cambridge, Cambridge University Press, 2006, pp. 301–318.

CHAPTER 3

Medical and Astrological Plates: Their Roles in Medieval and Renaissance Knowledge

Josefina Rodríguez-Arribas

This chapter concerns two plates in metal and two in paper, each made in Germany in the sixteenth century and used in two contexts, astrology and medicine. Its purpose is to describe the plates in order to understand how they might have been used and to find out more about the practitioners (astrologers or physicians) who might have been interested in producing, owning, and using these objects.*

Astrological and Astrological-medical Plates in Metal

The existence of medieval astrolabes with astrological engravings in several languages (Arabic, Latin, Hebrew, English, etc.) provides evidence for the involvement of astrologers with instruments in this period. The astrological information these instruments display includes the divisions of the houses on the plates, and diagrams and tables usually engraved on the backs of the astrolabes with the triplicities, aspects, terms, decans, symbols of the planetary rulers of some of these divisions, and the lunar mansions.[1] Some retes (the component showing the ecliptic and the positions of the fixed stars) also indicate the nature of some of the fixed stars, with engravings of the symbols of one or two planets that characterize the nature and influence of the specific star. One of the systems of house division is often represented on the latitude

* I wish to thank Thomas Eser, Roland Schewe and Anna-Lena Krämer of the Germanisches Nationalmuseum in Nuremberg for making the instruments and engravings available for study and for kindly sharing their knowledge and experience. I also thank Darrel Rutkin for his reading. This article was made possible by a Fellowship of the International Consortium for Research in the Humanities, "Fate, Freedom and Prognostication, Strategies of Coping with the Future in East Asia and Europe" (supported by the Federal Ministry of Education and Research) at the Friedrich Alexander Universität, Erlangen-Nuremberg, Germany.

1 Silke Ackermann, "Astrological Scales on the National Maritime Museum Astrolabes", in Koenraad Van Cleempoel, *Astrolabes at Greenwich: A Catalogue of the Astrolabes in the National Maritime Museum*, Oxford, Oxford University Press/National Maritime Museum, 2006, pp. 73–90.

 | DOI:10.1163/9789004381438_005

plates of planispheric astrolabes, while two other systems are easily calculated using the divisions of the seasonal hours that are often engraved on astrolabes.

Astrolabes are not, however, the only instruments that display information related to astrology or astrological medicine. The first part of this chapter discusses two metal plates that are not astrolabes and do not display the same astronomical complexity, but are designed for use in astrology and medicine.[2] Although both were made in the sixteenth century, they represent practices and theories that were mainstream by the Middle Ages.[3]

The most complex of the two plates (*Fig. 3.1*) is anonymous but dated.[4] It is an astrological plate probably made in Nuremberg, the free imperial city that became the main centre for astrolabe production in the fifteenth century. The instrument consists of three brass components that are interconnected: a flat plate below, a counterchanged rule in the middle, and a simplified rete on the top. The instrument also has a small and plain suspension ring, which must have been added later. The back is blank, other than the date 1516 engraved in Hindu-Arabic numbers.

The front of the plate is graduated with the three circles that are standard on astrolabes: the circles of Capricorn, the equator and Cancer. Similarly, it has a meridian line and east-west line (or straight horizon), which are perpendicular to each other and intersect at the centre of the plate. The outer border of the plate has scales of the hours and the houses. The hour scale has 24 equal

---

2 They are described (one quite briefly) in the catalogue of the collection published by the Germanisches Nationalmuseum, see Gerhard Bott (ed.), *Focus Behaim Globus*, Nuremberg, Verlag des Germanischen Nationalmuseums, 1992, 2 vols.; v. 2, pp. 592 (no. WI 22), and 624–625 (no. WI 1808); see also the bibliography for the two instruments.

3 See any handbook of medieval astrology, for instance, Charles Burnett and Keiji Yamamoto, *Abu Mashar, The Abbreviation of the Introduction to Astrology: Together With the Medieval Latin Translation of Adelard of Bath*, London, Brill, 1994; and Charles Burnett, Keiji Yamamoto, and Michio Yano, *Al-Qabisi (Alcabitius): The Introduction to Astrology*, London, Warburg Insitute, 2004. For astrology in medicine, see, for instance: Hiro Hirai, "The New Astral Medicine", in Brendan Dooley (ed.), *A Companion to Astrology in the Renaissance*, London-Leiden, Brill, 2014, pp. 267–286; Roger Kenneth French, "Foretelling the Future: Arabic Astrology and English Medicine in the Late Twelfth Century", *Isis* 87, 3 (1996), pp. 453–480; and R.K. French, "Astrology in Medical Practice", in *Ancients and Moderns in the Medical Sciences: From Hippocrates to Harvey*, Aldershot, Ashgate Variorum, 2000, pp. 30–59; Cornelius O'Boyle, *Medieval Prognosis and Astrology: A Working Edition of the* Aggregationes de crisi et creticis diebus: *With Introduction and English Summary*, Cambridge, Welcome Unit for the History of Medicine, 1991; and Nancy G. Siraisi, *Medieval and Early Renaissance Medicine: An Introduction to Knowledge and Practice*, Chicago-London, University of Chicago Press, 1990.

4 Germanisches Nationalmuseum, Nuremberg, WI 22; overall diameter 166–167 mm, thickness 1–1.5 mm; rete: diameter 105 mm, thickness 1 mm.

FIGURE 3.1 Astrological plate for calculating the house divisions, unsigned, 1516 (Germanisches Nationalmuseum, Nuremberg, WI 22; Photo: M. Runge).

sections corresponding to the 24 equinoctial hours of day and night, each numbered clockwise from 1 to 12. Each hour has five-minute sub-divisions. The scale of houses is divided into twelve unequal divisions that correspond to the twelve astrological houses of the horizon for which the plate was made. One of these circular arcs, the one embracing the cusps of the first house and its opposite, the seventh, is the line of the horizon. Finally, the lower half of the plate contains twelve unequal divisions for the seasonal hours. The rule is fixed by a pin through the centre of the plate and the rete. One of its arms is longer and has a small circle engraved on its end. This is the part of the rule used as a pointer on the plate. There are no other engravings on the rule.

All three components of the instrument are interconnected but can move separately. The rete is a simplified form of astrolabe rete: it contains the zodiac circle or ecliptic but no pointers for the fixed stars. It has three radii that join at the centre, through which a pin was attached to another piece, now missing.

The missing part was possibly an alidade with hour divisions to tell the time using either of the pinnules as the gnomon of a sundial.[5] One of the radii is attached to the rule and to the centre of the plate with a pin.

The simplified rete consists of three graduated circles for the ecliptic or zodiac: one with divisions for every two degrees; one with divisions of ten degrees each; one with the names of the zodiac signs in Latin, most of them in abbreviated forms common in Latin astrolabes and manuscripts.[6] The segments of the twelve zodiac signs are unequal. This is a standard and correct feature also found on astrolabes, since the zodiac/ecliptic is a heavenly sphere projected onto a flat surface (the rete). The centre of the rete is placed 20 mm away from the centre of the plate and is thus eccentric.

This instrument was specifically made to find the astrological houses at any hour of day for any day of any year at a specific horizon that remains uncertain. From the lengths of the longest and shortest days on the plate, the implicit latitude should be about 41° north, which is far from Nuremberg.[7] All the horary information displayed on the instrument is there to facilitate this calculation, either in equinoctial or in seasonal hours. The horary divisions of seasonal and equinoctial hours engraved on the plate are there not to calculate time, which is impossible, for the instrument alone cannot provide the information required: it has no altitude scale and no almucantars to place the position of the Sun and find time using the solar degree or its nadir. Rather, the rising degree at the chosen time is the only coordinate needed for the instrument to function. When this degree of the rete is placed on the horizon, the cusps of each of the remaining houses are indicated by numbers around the circle of Capricorn. The cusps are determined by the degrees of the ecliptic aligned with each of the house divisions engraved on the plate. The house division method that these divisions display is the equatorial method of fixed boundaries, which became popular in the Renaissance and was often named after Regiomontanus. The seasonal hour divisions are there to make possible the calculation of the houses by two alternative methods: the hour line method (the fixed rete method in North's terminology), the most popular method for the division of houses (which was very old but attributed to Placidus in the Renaissance); and the so-called standard method (the rotating rete method in North's terms). Thus this

5 I thank Jan de Graeve for his insightful comments in relation to the possible missing component of this instrument, and Richard Kremer and Jim Bennett for their comments.

6 A distinctive feature of the script on the rete is that the capital letters N and S are written in reverse. Thomas Eser (personal communication) notes that this is characteristic of Gothic script and epigraphy, reflecting a taste for variety that encouraged the embellishment of scripts and inscriptions. Gothic numbers also feature, particularly the form of seven (Λ).

7 With thanks to Josep Casulleras for his comments on this point.

plate can be used to calculate the houses according to three methods.[8] The rule works simply as a pointer to help with the calculations.

This is certainly an instrument that could provide an astrologer with the basic layout of a horoscope for any moment with just a movement of the hand. It was an aid for astrologers, but was not autonomous: the astrologer also needed an ephemeris (astronomical tables of planetary positions) and tables for other astronomical parameters related to the different calculations. There is no complete astrological interpretation without the positions of the planets in the zodiacal signs (and thus in the houses). More importantly, an astrolabe or tables would have been essential for finding the rising degree, the only datum required to make the instrument work, but which it could not itself provide.

The second plate (*Fig. 3.2*) enlarges the scope of application of astrological instruments to include medicine.[9] It was presumably made in Prague in 1585: the name of the maker, Erasmus Habermel, and the date are engraved around the middle.[10] The golden brass plate has an original throne and suspension ring similar to those of a standard astrolabe. It is a beautiful sixteenth-century piece that might equally serve a court physician or astrologer.

The instrument consists of two plates, the smaller on top of the larger and attached by a pin (now missing). The two are now stuck together, although the smaller plate was intended to turn over the larger one. The different size of the plates creates two clear surfaces: the limb (on the large plate) and the central part (on the small plate). There is also a pointer to indicate the hours on the limb, although this no longer moves. The limb is divided into 24 equal

8 For an explanation of these methods, see John D. North, *Horoscopes and History*, London, Warburg Institute, 1986, pp. 59, 64 and 66; Edward S. Kennedy, "The Astrological Houses as Defined by Medieval Islamic Astronomers", *Anuari de filologia, Secció B, Estudis àrabs i islàmics* 2 (1996), pp. 535–578; Josep Casulleras, "The Instruments and the Exercise of Astrology in the Medieval Arabic Tradition", *Archives Internationales d'Histoire des Sciences* 63 (2013), pp. 517–540: 519–523.

9 Germanisches Nationalmuseum, Nuremberg, WI 1808; diameter 120 mm, maximum size with throne 142 mm, thickness of large plate 7–10 mm, thickness of the whole 19–22 mm. It is almost identical to an astrological-medical plate at the Germanisches Nationalmuseum (Ph Z 3524), which is unsigned and undated.

10 Habermel, a renowned maker whose life is quite poorly documented, was working in Prague from 1585. He was appointed instrument maker to Emperor Rudolf II in Prague in 1593. See Ernst Zinner, *Deutsche und Niederländische astronomische Instrumente des 11. bis 18. Jahrhunderts*, 2nd ed., Munich, C.H. Beck, 1967, pp. 329–346 and 681; Wolfgang Eckhardt, "Erasmus Habermel: Zur Biographie des Instrumentenmachers Kaiser Rudolfs II", *Jahrbuch der Hamburger Kunstsammlungen* 21 (1976), pp. 55–92; Eckhardt, "Erasmus und Josua Habermel: Kunstgeschichtliche Anmerkungen zu den Werken der beiden Instrumentenmacher", *Jahrbuch der Hamburger Kunstsammlungen* 22 (1977), pp. 13–74; "Erasmus Habermel", *Epact: Scientific Instruments of medieval and Renaissance Europe*, www.mhs.ox.ac.uk/epact/maker.php?MakerID=40 (accessed 11 April 2018).

FIGURE 3.2 Front of an astrological-medical plate for calculating the crises of an illness, by Erasmus Habermel, 1585 (Germanisches Nationalmuseum, Nuremberg, WI 1808; Photo: M. Runge).

sections corresponding to the 24 equal hours of day and night. They are numbered clockwise from 1 to 12 from top and bottom. Each hour is divided to 15 minutes. The small plate has a wind rose in the centre. A hole through the middle of the plate indicates that a pin and wedge, and perhaps also a rule for the back, are now missing.

Surrounding the wind rose is a square with angles that have been enlarged to include some text. Each is engraved with the crises of an illness starting from its inception: "Principium Morbi", "Crisis Prima", "Crisis Secunda", and "Crisis Tertia". Each side of this square is divided with arcs into seven sections. Four of these are numbered consecutively, from the first after the "Principium Morbi" (1) to the last before the "Principium Morbi" (16). The central section of each side of the square displays the inscription "INDICATIVA" (indicative). The second and sixth divisions of each side also have the inscription "Intercidens" (separating). The first, third, fifth and sixth divisions on each side bear the word "Vacui" (empty). Three lines radiate from the pointer (which is the position of the "Principium Morbi") of the small plate to the left and right to indicate the astrological aspects of sextile (60°), square (90°) and trine (120°). The purpose of the instrument is clearly indicated by the engraving "DIES MEDICINALES". Each letter of this phrase is engraved on each of the divisions labelled "Vacui", starting clockwise from the first division before "Principium Morbi".

The back of the instrument (*Fig. 3.3*) displays astrological information concerning crises and the phases of the Moon. A missing rule would have pointed to the different days and phases of the Moon, while another on the front could have pointed to the evolution of the illness according to the crises. The back consists of the three standard concentric circles (the circle of Cancer, the equator and the circle of Capricorn) and the perpendicular meridian and east-west lines. Twenty-four unequal divisions for the 24 hours are engraved with Hindu-Arabic numbers 1–12. "Dieis", "Oriens", "Noctis" and "Occidens" are engraved at the top, east, bottom and west respectively, with "Horae Inaequales" engraved around the centre.

Outside the circle of Capricorn there are five concentric circular scales:

1. equinoctial hours in Roman numerals, divided to 15 minutes;
2. the humours: "Phlegma" (top), "Cholera" (bottom), "Melancho[lia]" (east), "Sanguis" (west). The same scale includes the nature of the four seasons according to the prevailing humour;[11]
3. twenty-eight equal divisions starting at the top and numbered anti-clockwise to indicate the days of the lunar cycle, in Hindu-Arabic numerals;
4. the phases of the Moon, indicated with drawings in the four directions of the plate: new (top), crescent (east), full (bottom) and waning (west), and between them the Moon's nature between phases;[12]
5. the Moon's influence on the four ages of man.[13]

11 Between "melancho" and "phlegma": "autumnus senectus occidentali terra semi"; between "phlegma" and "sanguis": "hiemes decrepitus septentri[o]nali aqua semi"; between "phlegma" and "sanguis": "orientali masculini aer juventali"; and between "cholera" and "melancho": "in aestas virilibus merientali ignus".

12 Between new Moon and crescent: "Ab Usquae ad I □ Calida et Humida"; between crescent and full Moon: "a Prima □ ad ☍ Calida et Sicca"; between full and waning Moon: "a ☍ Usquae ad 2 □ Frigida et Sicca"; and between waning and new Moon: "[a] Secunda □ ad ☌ Frigida et Humida".

13 Between new Moon and crescent: "a ☌ in □ Primam Apta est Juvenibus"; between crescent and full Moon: "a I □ in ☍ Electa Juvenibus et Virilibus"; between full and waning Moon: "a ☍ in 2 □ Bona Virilibus et Senibus"; and between waning and new Moon: "a 2 □ Usquae ad ☌ est Felona Senibus".

FIGURE 3.3 Back of the astrological-medical plate, showing information related to the lunar cycle, by Erasmus Habermel, 1585 (Germanisches Nationalmuseum, Nuremberg, WI 1808; Photo: M. Runge).

Dividing the word "Dieis", there is a perpendicular zodiac scale with the symbols of the zodiac signs in pairs in sections of unequal size according to their diurnal arcs: Gemini/Cancer; Taurus/Leo; Aries/Virgo; Pisces/Libra; Aquarius/Scorpio; Capricorn/Sagittarius. This division of the zodiac by the axis of the sosltices (in Cancer and Capricorn) into two parts shows the relation between every two degrees of the zodiac that are equidistant with respect to either of the two sosltices. The days in which the Sun is in either of these two degrees, its position in the sky and the length of the daylight in they both are equivalent.

The purpose of this instrument is to calculate the critical days (crises) of an illness. These are determined by the lunar cycle of 28 days, so that every seven days there is a change (*crisis prima, crisis secunda,* and *crisis tertia*) in the illness, for better or worse. According to this doctrine, the fourth day of each of these critical hebdomadas ("Indicativa" on the plate) is indicative of the outcome of the coming crisis.[14] This is a lunar instrument, therefore, in that the information it displays relates exclusively to the Moon and its influences. It is not an instrument of calculation, but more likely an instrument for teaching or for displaying during medical practice as a symbol of learning. During the Middle Ages and Renaissance, a physician who also mastered astrology excelled in his profession, so physicians might well include astrological terminology in their technical speech, even if their knowledge of the technicalities of practical astrology could be often superficial.

These two metal plates were neither observing instruments nor fully calculating ones. The second is likely related to the royal court of Rudolph II of Prague and can be called a "court instrument", the product of a specific social and political context that characterized many Renaissance cities (and certain Arabic and Christian courts of the Middle Ages). Both instruments were designed and made for people who might not have had a detailed knowledge of astrology, but were interested in displaying familiarity with it for reasons related to power, prestige and legitimation. The additional (but no lesser) value of using instruments for such things as medicine and divination should not be underestimated in a period that was fascinated with mechanical objects, machines and the inventions of engineers.[15]

---

14 For the theory of the critical days in ancient and medieval sources, see: Tzvi Y. Langermann, "Critical Notes on a Study of Galen's On Critical Days in Arabic, or A Study in Need of Critical Repairs", *Aestimatio* 9 (2012), pp. 220–240; Glen M. Cooper, *Galen, De diebus decretoriis, from Greek into Arabic: A Critical Edition, with Translation and Commentary of Ḥunayn ibn Isḥāq's Kitāb ayyām al-buḥrān*, Farnham, Ashgate, 2011, pp. 61–76; Concetta Pennuto, "The Debate on Critical Days in Renaissance Italy", in A. Akasoy, C. Burnett and R. Yoeli-Tlalim (eds.), *Astro-Medicine: Astrology and Medicine, East and West*, Florence, SISMEL: Edizioni del Galluzzo, 2008, pp. 75-98; Danielle Jacquart, "Le temps médical au moyen âge, ou l'introuvable précision", *Bibliothèque de l'École des Chartes* 157, 1 (1999), pp. 157–170; Giuseppe dell'Anna, *Dies Critici: La teoria della ciclicità delle patologie nel XIV secolo*, Galatina, Mario Congedo Editore, 1999, 2 vols.; v. 1, pp. 43–82; Ursula Weisser, "Mondphasen und Krisen nach al-Kindī", *Sudhoffs Archive für Geschichte der Medizin und der Naturwissenschaften* 66 (1982), pp. 390–395; Felix Klein-Franke, "Die Ursachen der Krisen bei akuten Krankheiten: eine wiederentdeckte Schrift al-Kindi's", *Israel Oriental Studies* 5 (1975), pp. 161–188.

15 See for instance: Darin Hayton, "Instruments and Demonstrations in the Astrological Curriculum: Evidence from the University of Vienna, 1500–1530", *Studies in History and Philosophy of Biological and Biomedical Sciences* 41 (2010), pp. 125–134; Jonathan Sawday, *Engines of the Imagination: Renaissance Culture and the Rise of the Machine*, Abingdon, Routledge, 2007, pp. 83–86 and 96–102.

Both instruments are composite: they consist of different components that work together to attain a result. In the case of the first plate, the result sought is eleven different degrees of the ecliptic that mark the cusps of eleven of the twelve astrological houses. As discussed, the cusp of the first house must be found by other means and is essential for the instrument to work. For the second plate, the result sought is a set of specific days of the lunar month that relate directly to the beginning and course of a patient's illness. The first plate is a functional/working scientific instrument (a calculator); the second is not so much an instrument as a visual aid. There are graduated scales, but the instrument does not calculate anything. It displays information in a diagrammatic form that helps the physician understand how critical days work, but he must find the specific dates of the stages of the illness separately. The plate does not display this information, just the general theoretical scheme. It remains intriguing that this plate has scales to tell the time in both seasonal and equinoctial hours, yet has nothing for calculating the Sun's altitude and using it with the local latitude to tell the time. Either the horary scales were just a reminder and a scientific motif with no practical application, or there was a missing alidade with graduations for use as a sundial.

Astrological and Astrological-medical Plates in Paper

Two engravings preserved at the Germanisches Nationalmuseum in Nuremberg (*Figs. 3.4* and *3.6*), both made by Georg Hartmann in Nuremberg in 1554, are strikingly similar to the brass plate made by Erasmus Habermel. Hartmann (1489–1564) was an instrument maker who also wrote treatises on instruments. He settled in Nuremberg in 1518 and shortly before 1523 set up a workshop specializing in the production of scientific instruments, of which more than thirty survive.[16]

16 See "Georg Hartmann", *Astronomie in Nürnberg*, www.astronomie-nuernberg.de/personen/show.asp@ID=45 (accessed 11 April 2018), and "Geräte von Georg Hartmann", *Astronomie in Nürnberg*, www.astronomie-nuernberg.de/index.php?category=hartmann&page=astrolabien (accessed 11 April 2018). Regarding Hartmann and his instruments, see British Museum, "Georg Hartmann", www.britishmuseum.org/research/search_the_collection_database/term_details.aspx?bioId=85909 (accessed 11 April 2018); John P. Lamprey (ed. and tr.), *Hartmann's Practika: A Manual for Making Sundials and Astrolabes with Compass and Rule*, Bellvue CO, John Lamprey, 2002; Lamprey, "An Examination of Two Groups of Georg Hartmann Sixteenth-century Astrolabes and the Tables Used in their Manufacture", *Annals of Science* 54 (1997), pp. 111–142; H.G. Klemm, *Georg Hartmann aus Eggolsheim (1489–1564): Leben und Werk eines fränkischen Mathematiker und Ingenieurs*, Forchheim, Ehrenbürg-Gymnasium Forchheim, 1990, pp. 77–83; E. Zinner, *Deutsche und*

FIGURE 3.4 Plate displaying astrological and medical information, by Georg Hartmann, 1554 (Germanisches Nationalmuseum, Nuremberg, HB 2713; Photo: G. Janßen).

The first engraving is of a medical-astrological plate displaying copious information (*Fig. 3.4*).[17] It consists of a set of concentric circular scales and polygons:

1. the cardinal directions;
2. the zodiac, divided to 5° and 1°;
3. the rulers of the triplicities (fire, air, water and earth) for each zodiac sign;[18]

---

niederländische astronomische Instrumente des 11.-18. Jahrhunderts, München, C.H. Beck, 1979, pp. 357–368.

17 Germanisches Nationalmuseum, Nuremberg, HB 2713. Another copy is at the Bayerische StaatsBibliothek, München: Georg Hartmann: [Collectio figurarum], *Eine reiche Sammlung von Kupferstichen u. Holzschnitten, welche Sonnenuhren u. andere astronomische Instrumente darstellen*, S.l., [16. Jh.], Rar 434, fol. 102v. Dimensions: 240 mm (diameter).

18 There is a mistake in the rulers of the triplicity of Sagittarius (fire), which are written Sun-Saturn-Jupiter, but should be Sun-Jupiter-Saturn (as they are in Leo and Aries).

4. the dignities of the planets (domicile, exile, exaltation and detriment);
5. melothesia (the parts of the body ruled by each zodiac sign);[19]
6. the 28 mansions of the Moon, each labelled with a capital letter (T = *temperata*; S = *sicca*; H = *humida*), the first at about 19.5° of Aries:

1	2	3	4	5	6	7	8	9	10	11	12	13	14
T	S	H	H	S	T	H	T	S	H	T	H	T	T
15	16	17	18	19	20	21	22	23	24	25	26	27	28
H	H	H	S	H	H	T	H	T	T	S	S	H	T

7. the lunar cycle of 27½ days, with subdivisions of six hours for each day;
8. the qualities of the lunar phases;[20]
9. the critical and indicative days with the periods that separate them;
10. the astrological aspects: sextile, square, opposition and trine.

The most specific information pertains to medicine, namely the relative positions of the Moon and the critical days:

Stage in the disease	**Days. Hours. Minutes**
Periodus	1. 16. 59.
1 Indicativus	3. 1. 58.
Periodus	5. 2. 57.
Primus Creticus	6. 19. 56.
Periodus	8. 12. 55.
2 Indicativus	10. 5. 54.
Periodus	11. 22. 53.
Secundus Creticus	13. 15. 52.
Periodus	15. 8. 50.
3 Indicativus	17. 1. 49.
Periodus	18. 18. 48.
Tertius Creticus	20. 11. 47.
Periodus	22. 4. 46.

19 Aries: *caput, vultum, pupilla*; Taurus: *collum, artherias, vocis*; Gemini: *spatulas, brachia, manus*; Cancer: *pectus, stomachum et pulmonem*; Leo: *cor et costas*; Virgo: *ventrem et lumbos*; Libra: *anchas et nates*; Scorpio: *verenda*; Sagittarius: *crura supra tybias*; Capricorn: *tybias ipsas*; Aquarius: *crura sub tybias*; and Pisces: *cavillas et pedes*.

20 *Prima quadra lun[a]e calida et humida terminat humidum; Secunda quadra lun[a]e calida et sicca terminat calidum; Tert[i]a quadra lun[a]e frigida et sicca termina[t] siccum; Quarta quadra lun[a]e frigida et humida terminat frigidum.*

Table (*cont.*)

Stage in the disease	Days. Hours. Minutes
4 Indicativus	23. 21. 45.
Periodus	25. 14. 44.
Quartus Creticus	27. 7. 45.

According to the aspect diagrams, there is always an aspect of square (90°) between two critical days and semi-square (45°) between a critical day and the indicative. The numbers of the second column are the days, hours and minutes separating each of the periods, indicative days and critical days. Each period ("periodus") stands for days that are indifferent as regards their influence. Only indicative ("indicativus") and critical ("creticus") days are significant for the patient.

This *directorium medicorum* or physicians' guide is thus a visual tool to help understand how any illness should evolve according to the astrological aspects of the Moon with the position of the Moon at the inception of the illness. However, this is just general information; a physician eager to apply astrology to his practice might also consult astrological handbooks describing the influences of the planets and their aspects, also taking into consideration the patient's natal horoscope. This technical information, which is essential for medical astrology, does not appear on the plate. As with the two metal plates, this begs the question of whether a physician using the plate would have had a professional knowledge of astrology. Knowledge of the dignities and rulers of the triplicities would be basic for any student of astrology and does not help in understanding how to interpret the many components of an astrological figure and their interrelations. It seems that certain physicians of the sixteenth century (as in earlier centuries) concentrated on a limited range of astrological factors: the Moon's cycle and its influence on critical days, the influence of the circadian and annual cycles of the Sun on the humours, and the rulership of the zodiacal signs over the different parts of the body. Other factors (aspects between the planets, directions, transits, horary astrology, etc.) might be ignored.

It is worth noting the existence of a similar but earlier paper plate (another *directorium medicorum*) by Georg Hartmann, dated 1538 (*Fig. 3.5*).[21] It is not

21 Wolfenbüttel, Herzog-August-Bibliothek, 15 Astron. 2°, 1a. See also Suzanne Karr Schmidt, "Art—A User's Guide: Interactive and Sculptural Printmaking in the Renaissance", PhD,

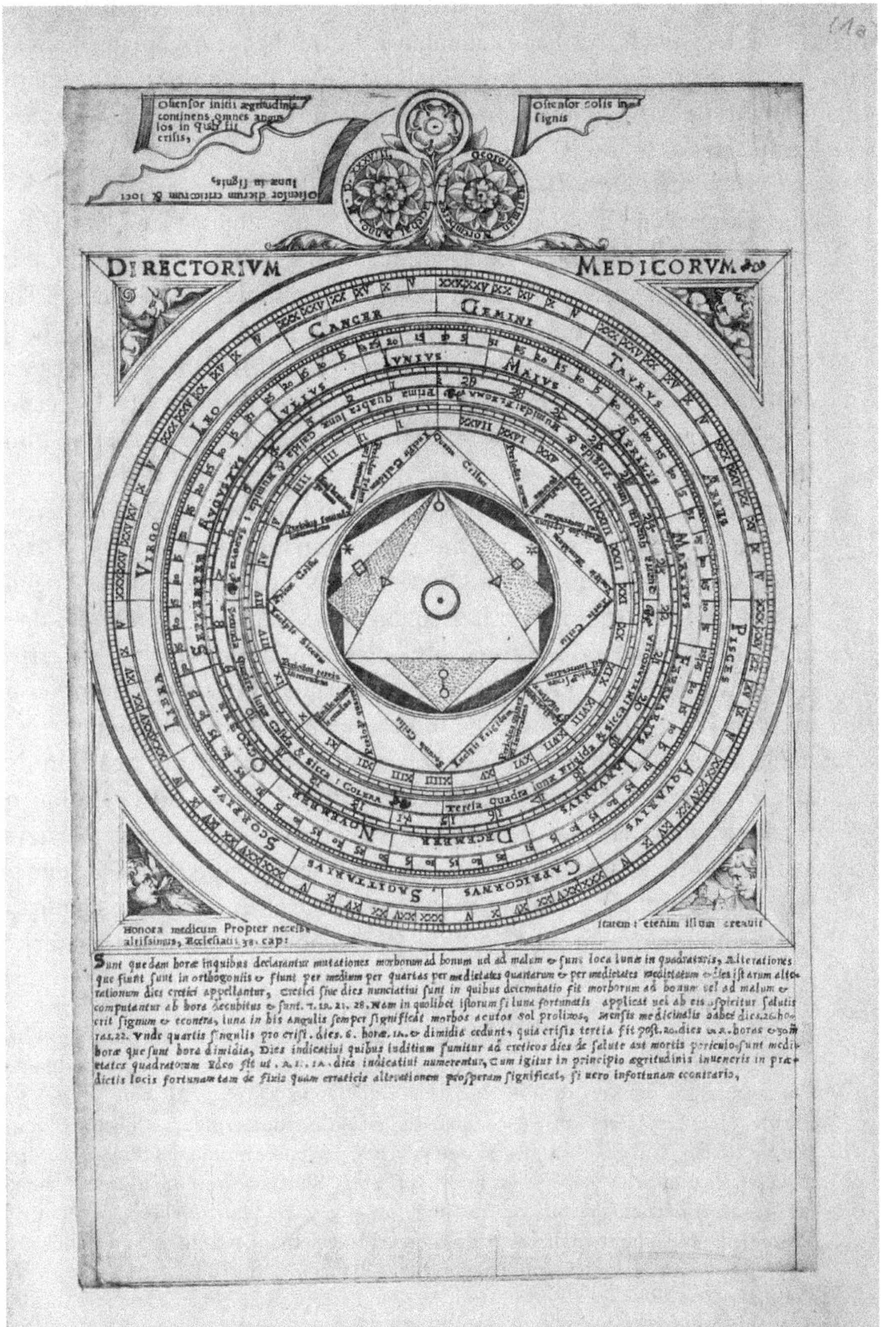

FIGURE 3.5 Astrological-medical engraving with information about the crises, by Georg Hartmann, 1538 (Herzog-August-Bibliothek, Wolfenbüttel, 15 Astron. 2°, 1a).

identical, however, since it lacks the scales with the rulers of the triplicities and dignities of the planets, but has a calendar scale. Unlike the 1554 plate, it has an explanatory text that seems to be a modification of a chapter on iathromathematics by the physician and astrologer Girolamo Manfredi (1455–1492) and which translates as follows:

> There are certain hours in which changes in any illness take place either for good or for bad and are determined by the positions of the Moon in square [with respect to its position at the beginning of the illness]. The alterations happen at [certain] angles: opposition [180°], square [90°], semi-square [45°], and their halves, and the days of these changes are called critical days. Critical or indicative days are those in which one determines whether the illness gets better or worse, and are counted from the hour at which the person [starts] lying [in bed]. These days are the 7th, 14th, 21st and 28th because it is a sign of health if the Moon joins the fortunate [planets] or receives an aspect from them in any of these days. By contrast, the Moon in these angles always means serious and long illnesses. The medical month has 26 days and 22 hours, therefore 6 days and 17½ hours correspond to every quarter [of a month] that is a crisis. In this way, the third crisis falls after 20 days, 4 hours, and 30 minutes of an hour [i.e. half an hour]. The indicative days, in which a judgment is made as regards the [recovery of] health or the danger of death in the critical days, are placed in semi-square [with respect to the beginning of the illness], so that the 4th, 11th, and 17th are considered indicative days. Consequently, if you find a fortunate [star], either from the fixed stars or the planets in these positions at the beginning of the illness, it signifies a favourable change, but the opposite [if you find] an unfortunate [star].[22]

Yale University, 2006, pp. 180–273. Another astrological plate by Georg Hartmann has not been studied for this chapter: Astrologische Scheibe aus Messing, Georg Hartmann, 1541, formerly in the Kunsthistorisches Museum, Vienna, KK 9823, AR Kat.-Nr. 1536; see Klemm, op. cit. (n. 15), pp. 28, 83. It was auctioned at Christie's, London, 8 July 1999, lot 177 (Paulus Rainer, Kunsthistorisches Museum, Wien, personal communication). According to a conversation with Stephen Johnston, this brass plate is now in the Museo Nacional de Ciencia y Tecnología, Madrid, inv. 1999/025/0003, see Marcos Villaverde Aparicio, "Astrología y religion: Georgius Hartmann en el Museo Nacional de Ciencia y Tecnología", in Josep Batlló Ortiz, Jordi Ferran Boleda, Mercè Piqueras Carrasco (eds.), *Actes de les VIII Trobades d'Història de la Ciència I de la Tècnica*, Barcelona, Societat Catalana d'Història de la Ciència i la Tècnica, 2006, pp. 299–306. I thank S. Johnston for this update and the bibliographical reference.

22 Translation by the author. Original Latin: "Sunt quedam horae inquibus declarantur mutationes morborum ad bonum vel ad malum et sunt loca lunae in quadraturis, alterationes que fiunt sunt in orthogoniis et fiunt per medium per quartas per medietates quartarum

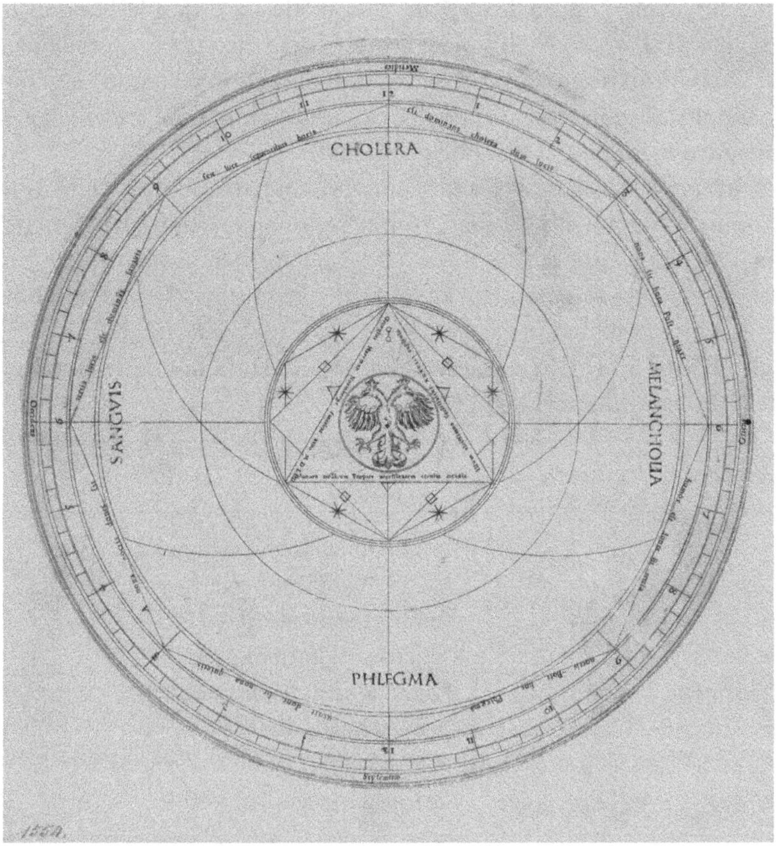

FIGURE 3.6 Medical plate showing temperaments, horary divisions and astrological aspects, by Georg Hartmann, 1554 (Germanisches Nationalmuseum, Nuremberg, HB 25492; Photo: M. Runge).

---

et per medietates medietatum et dies istarum alterationum dies cretici appellantur, cretici sive dies nunciativi sunt in quibus determ[i]natio fit morborum ad bonum vel ad malum et computantur ab hora decubitus et sunt 7, 14, 21, 28. Nam in quolibet istorum si luna fortunatis applicat vel ab eis aspicitur salutis erit signum et econtra, luna in his angulis semper significat morbos acutos sol prolixos. Mensis medicinalis habet dies 26. horas 22. unde quartis singulis pro crisi dies 6 horae 17 et dimidia cedunt, quia crisis tertia fit post 20 dies et 4 horas et 30 m[inuta] horae que sunt hora dimidia. Dies indicativi quibus iuditium sumitur ad creticos dies de salute aut mortis periculo, sunt medietates quadratorum ydco fit ut 4, 11, 17 dies indicativi numerentur, cum igitur in principio aegritudinis inveneris in praedictis locis fortunam tam de fixis quam erraticis alterationem prosperam significat, si vero infortunam econtrario." Hartmann's version is more convoluted than the original in Girolamo Manfredi, *De medici et infirmis collectanea in ordinem centiloquii congesta*, Nuremberg, Georg Wachter, 1530, Verbum LXXVII: "In morbis quidem sunt dies et horae, quibus declarantur mutationes eorum ad bonum vel ad malum[q]ve lociter. Et

The second engraving is a medical plate also made by Georg Hartmann in Nuremberg in 1554 (*Fig. 3.6*).[23] This plate is more basic than the one previously described: the information concerns only humoral theory related to the hour of the day and the astrological aspects, all of it based in medieval knowledge. It consists of concentric circular and polygonal scales:

1. the four main directions of the horizon that divide the plate into four equal quadrants: "Meridies" (top), "Septemtrio" (bottom), "Oriens" (right), "Occidens" (left);
2. the equal hours numbered clockwise from 1 to 12 from the top and bottom, each hour divided to 15 minutes;
3. a scale divided into eight equal sections with the humour corresponding to each three-hour interval:

Hours	Humours
12–3 (from the top)	"est dominans cholera dum lucis"
3–6	"nona sit hora post niger"
6–9	"humor est donec sit tercia"
9–12	"noctis post hoc phlegma"
12–3 (from the bottom)	"venit donc sit nona quietis"
3–6	"a nona noctis donec sit"
6–9	"tertia lucis est dominans sanguis"
9–12	"sex luce sequentibus horis"

4. the circles of Capricorn, Cancer and the equator (projected as is standard on planispheric astrolabes), divided into four quarters; the ends of

sunt dies cretici seu pugnae, et hoc erit, cum Luna ad quadraturam vel oppositum sui loci initii aegritudinis pervenerit. Unde fit, ut septimus, decimusquartus et vigesimus vel vigesimusprimus dies cretici a medicis appellentur. Dies autem iudicativi, quibus iudicium sumitur, ad creticos dies de salute aut morte, mortis[q]ve periculo, sunt medietates quadratorum, ideo fit, ut quartus, undecimus, et decimusseptimus dies iudicativi numererentur. Et hic accipitur quadratura vel aspectus per circulum directum, et non per gradus aequales, et hoc erit, si processerit aegritudo secundum qualitatem naturalem, et non acaderit error, qui disturbet infirmum. Cum igitur in principio aegritudinis inveneris in praedictis locis fortunam, tam de fixis, q[uam] de erraticis, alterationem prosperam significabit. Si vero infortunam, alteratione[m] adversam, nisi fuerit ipsa infortuna aegritudini co[n]traria, et fortuna ei similis. Et Luna in his angulis semper significat morbos acutos et sol prolixos."

23 Germanisches Nationalmuseum, Nuremberg, GNM HB 25492. Dimensions: 240 mm (diameter). Another copy is at the Bayerische StaatsBibliothek, München: Hartmann, Georg: [Collectio figurarum], *Eine reiche Sammlung von Kupferstichen u. Holzschnitten, welche Sonnenuhren u. andere astronomische Instrumente darstellen*, S.l., [16. Jh.], Rar 434, fol. 71v.

the dividing lines (which are the cardinal directions for this specific horizon) are labelled: "CHOLERA" (midday), "MELANCHOLIA" (6 pm), "PHLEGMA" (midnight) and "SANGUIS" (6 am);

5. a scale that divides the inner surface of the plate between the circles of Capricorn and Cancer into eight unequal and unnumbered parts;
6. diagrams for the astrological aspects, drawn within the circle of Cancer: hexagon (sextile, 60°), square (square, 90°), triangle (trine, 120°), diameter (opposition, 180°);
7. a crowned double eagle inscribed within the triangle, along the sides of which is an inscription; the double eagle has three small circles (black, white, and grey) on its front.[24]

The plate relates the hour of the day to a specific humour. This could be an indication that it was intended to display information about the best time to perform a specific task according to the dominant humour (a sort of electional theory not based directly on technical astrology but on medicine, and more specifically on the analogy between the macrocosm and the microcosm as reflected in the humours and the daily solar cycle). In contrast to the previous examples, the layout of the plate is more cosmological than astrological, with its applications going beyond astrology and medicine. It seems most likely that it was either for display as a visual aid for choosing a propitious time or, more likely, an aide-mémoire for teaching humoral theory.

Conclusion

The anonymous astrological plate in metal discussed in the first part of this chapter cannot be used for observation in its present state and has clear limitations for use as a calculator. It enables one step in a longer series of calculations that involved the use of other instruments or tables, which could account for its limitations as a calculating device. The Habermel medical plate simply displays general information that the user would have to re-elaborate with other tools to apply it to a specific person or circumstance. Both instruments are manifestations of the increasing popularization of some aspects of the sciences (astronomy, medicine, alchemy, etc.) that began in the Renaissance and became apparent in iconographic motifs in the period.

24 The inscription reads: [base] "honora medicum propter necessitatem et enim creavit"; [right] "illum altissimus Ecclesiatici XXXVIII capitulo"; [left] "Georgius Hartman Normberg[a]e faciebat anno MDLIIII". The double-headed eagle was a symbol of the Holy Roman Empire originating in the fifteenth century. It was a prominent motif in Nuremberg.

As for the paper plates discussed in the second part of this chapter, it is well known that in the Renaissance models of instruments in paper were given to craftsmen for reproduction in metal.[25] One can picture a workshop involving several workers, each with specific skills (metalworkers, astronomers, mathematicians, etc.). Georg Hartmann worked in the context of one of these workshops. He produced paper models (of astrolabes, latitude plates and sundials that survive) as a first step in the production of instruments in metal. The engravings could be reused many times and became in themselves works of art and skill that were valuable in their own right and were frequently collected.

Paper instruments could also be cheaper, more easily producible versions either of metal instruments or of some of their components. During the Middle Ages and Renaissance, the rete and latitude plate/s of astrolabes were sometimes made on paper and then pasted onto wood to make them more durable. More often, those parts of the instrument that were more difficult to make in metal (e.g. the rete with the intricacies of the star pointers) were made in paper and placed directly onto the surface of a metal mater or latitude plate without pasting them onto any other material. The evidence of these practices comes from Renaissance instruments made of paper and from late-medieval manuscripts, which referred to them in explanations of construction, although no medieval instruments or components of medieval instruments survive in paper with clear signs of having been used.[26]

However, the paper plates considered in this study are instruments of neither observation nor calculation. Rather, they display cosmological and medical theories in a complex visual diagram that interrelates them at different levels according to the law of analogy (between microcosm and macrocosm). These plates might have had corresponding versions in metal, like the metal medical plate by Habermel, but they remain purely visual tools rather than scientific instruments or calculating devices.

---

25 For this use of the diagrams, see Catherine Eagleton, "Medieval Sundials and Manuscript Sources: The Transmission of Information about the Navicula and the Organum Ptolomei in Fifteenth-Century Europe", in Sachiko Kusukawa and Ian Maclean (ed.), *Transmitting Knowledge: Words, Images, and Instruments in Early Modern Europe*, Oxford, Oxford University Press, 2006, pp. 41–71; Suzanne Karr Schmidt and Kimberly Nichols, *Altered and Adorned: Using Renaissance Prints in Daily Life*, Chicago, Art Institute of Chicago, 2011, pp. 73–82; Samuel Gessner, "The Use of Printed Images for Instrument-Making at the Arsenius Workshop", *Early Science and Medicine* 18.1–2 (2013), pp. 124–152.

26 See for instance, Josef Ṭaiṭaṣaq's explanation of how to construct an astrolabe in which a rete in paper is crafted to be used onto a latitude plate in metal, Ms. Oxford, Bodleian Reggio 47, fol. 15b (Hebrew, copied in Mantua in 1499). An edition of this manuscript is currently in preparation.

CHAPTER 4

"Preciseness and Pleasure": The Astrological Diptychs of Thomas Hood

Stephen Johnston

This paper focuses on two manuscript astrological diptychs by Thomas Hood, best-known now as a mathematical lecturer and author of late sixteenth-century London. While the doctrine that these large diptychs embody is introductory and conventional, their format is novel. The arrangement of material and its exquisite presentation together mark a distinctive contribution to the genre of astrological instruments.

I will describe the devices and their use, situating them within the specific context of Hood's career and suggesting their likely audience. I also seek their sources, identifying the contemporary printed works and devices Hood drew on for the content of his diptychs. While they provide much information in compendious form, they are not just intellectual objects. Examining their artful form and decoration provides new evidence for the international dissemination and reception of Renaissance astrolabe designs.*

Description and Use

Each diptych consists of two sheets of illuminated parchment mounted on a hinged pair of oak boards, each board being 43 by 43 cm. The overall format is the same: one leaf carries the mater and rete of an astrolabe and the other a comprehensive astrological table, arranged within a zodiac circle with rotating index and volvelle. While there are differences in their detailed form and layout, the diptychs represent two different versions of the same device.[1]

* Thanks to Andrea Clarke and Sue Ashpitel at the British Library for their help and to Silke Ackermann, then of the British Museum, who interceded with the British Library Manuscripts Department back in 1998 to arrange the reconfiguration of the diptychs' moving parts.

1 The diptychs have not previously been subjected to detailed study. I first saw them in 1997 and presented a preliminary report at the XVII Scientific Instrument Symposium in Sorø, Denmark, in 1998. The text of my spoken version was put online in about 2005 (www.mhs.ox.ac.uk/staff/saj/hood-astrology/) in the hope that others with more astrological expertise than myself would develop it further. This online presence succeeded in drawing some

 | DOI:10.1163/9789004381438_006

FIGURE 4.1 Double spread of astrological diptych, by Thomas Hood (British Library Board, Add. MS 71494).

The two items are now held by the British Library. The simpler but more complete of the two is Additional MS 71494 (*Fig. 4.1*).[2] On the left, the astrolabe limb consists of a neatly divided and coloured hour scale surrounding a mater whose most prominent feature is the set of houses constructed according to the Regiomontanus method. The circular arcs are marked in dark ink and each house contains a Latin text describing its character. The horizon, representing the cusp of the first and seventh houses, is labelled at its eastern and western ends in red. There is no equator but the Tropic of Cancer is drawn; above the horizon it appears as a dotted dark line, below it is coloured red and marked with the numbers of the unequal hours whose arcs terminate there. The unequal hour lines are themselves coloured red to distinguish them from the houses. The mater carries neither almucantars nor azimuths.

Within the Tropic of Cancer is a circular table of the planetary rulers of the 24 day and night hours, which is read with the help of the simplified rete that rotates above it. The pasteboard rete has eight asymmetrically arranged

attention to its subject (see Deborah E. Harkness, *The Jewel House: Elizabethan London and the Scientific Revolution*, New Haven and London, Yale University Press, 2007, p. 133), but did not prompt a deeper study.

2 High resolution images of all accessible surfaces of both diptychs were made available online on the British Library website in 2017: www.bl.uk/manuscripts/Viewer.aspx?ref=add_ms_71494_f001r and www.bl.uk/manuscripts/Viewer.aspx?ref=add_ms_71495_f001r (accessed 6 April 2018). These provide much more detail than can be reproduced in print.

spokes marked to act as indices for the underlying table: on each spoke is the sequence of seven planetary symbols of the days of the week, reading inwards from Sunday to Saturday. The rete's only celestial element is the ecliptic, with zodiac signs given by both name and symbol. The visual effect of the astrolabe is enhanced by a decorative scheme of blue, white and red to pick out degrees and other elements.

The facing sheet consists of a large circular table with a volvelle and rotating index, providing a comprehensive astrological and medical synopsis of the zodiac (*Fig. 4.2*). The table consists of several concentric sections separated by thin decorative bands of blue, white and red. The outermost section of the table is the zodiac divided into signs and degrees. Within this is a section of twelve circular scales displaying the essential dignities and debilities of the planets in the signs. The rotating index rule conveniently identifies each scale, reading inwards: faces or decans ("Facies"); terms ("Termini"); triplicities ("Triplicitat[es]"); exaltations ("Exaltatio[nes]"); mansions or domiciles ("Domus"); rejoicing ("Gaudia"); a scale combining both depression or fall and detriment ("Casus et detriment[um]"); masculine and feminine degrees ("Gradus masc[ulini] et fœminini"); lucid, dark, empty and smoky degrees ("Lucidi tenebr[osi] vacui fum[osi]"); pitted ("Puteales"); lame ("Azemeni"); and degrees increasing fortune ("Augentes fortunam").[3]

The next section shows the parts of the body ruled by the planets depending on the zodiac sign they occupy ("Quibus corporis partibus quilibet planeta in quovis signo præsit"). It represents a step beyond the familiar Zodiac Man, consisting of seven rings, each identified by a planetary symbol from Saturn inwards to the Moon, with a corresponding body part named for each zodiac sign. The last section of the table provides a descriptive account of each of the zodiac signs and their influence on personal character and health ("Signorum variæ qualitates"). It surrounds a ring which reiterates the sequence of zodiac signs for ease of reference.

Within the table is a final functional element of the sheet, a central volvelle for critical days, again identified on the rotating index. The disc is divided to sixteen points and uses colour and differing numeral types to distinguish between four crises, eight "intercidens" days and four "indicativus" days. There is a projecting index for the onset of disease ("Ineuntis morbi index").[4]

---

3 See J.C. Eade, *The Forgotten Sky: A Guide to Astrology in English Literature*, Oxford and New York, Clarendon Press, 1984, pp. 59–73, for the character and interrelation of these qualities.

4 For an introduction to the complex and controverted doctrine of critical days in the Renaissance, see Glen M. Cooper, "Approaches to the Critical Days in Late Medieval and Renaissance Thinkers", *Early Science and Medicine* 18 (2013), pp. 536–565.

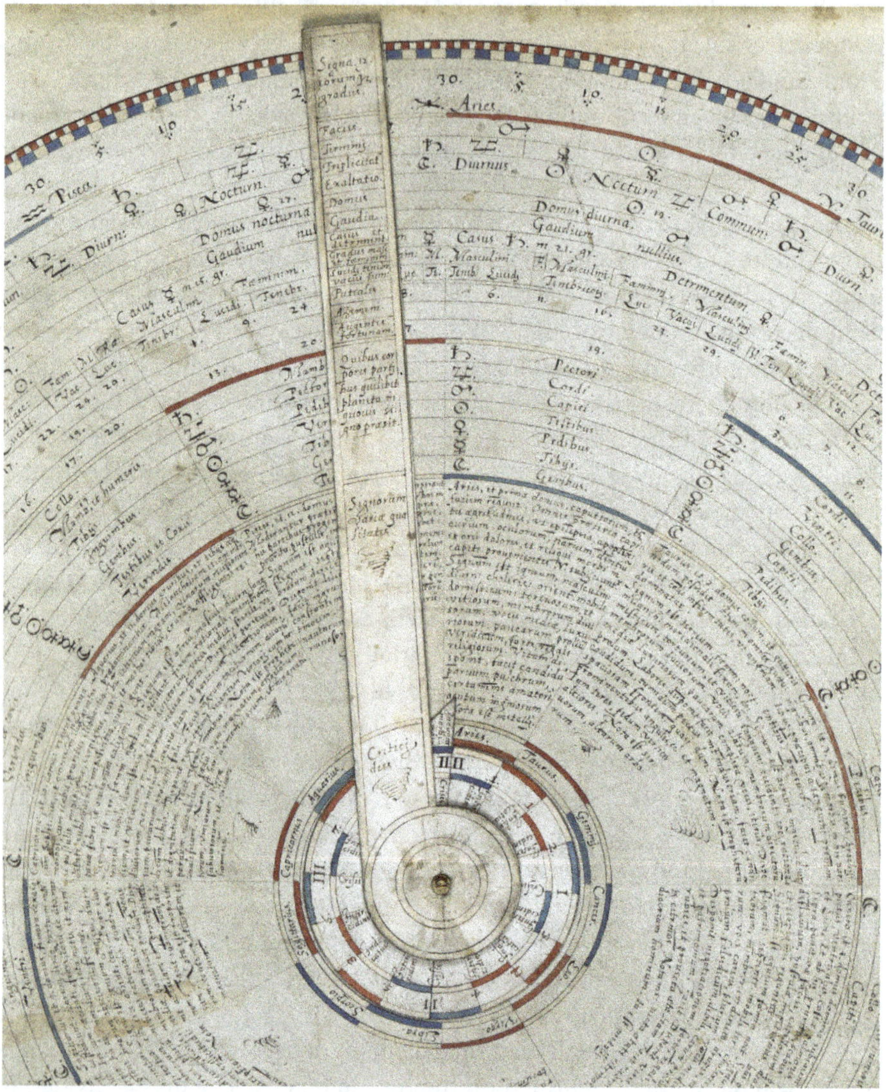

FIGURE 4.2 Detail of Thomas Hood's astrological table showing the rotating index and volvelle of critical days (British Library Board, Add. MS 71494).

The second diptych can be described much more quickly. Additional MS 71495 (*Fig. 4.3*) contains the same elements, with some rearrangement and minor disparities in detail (for example, this astrolabe's limb carries degrees as well as hours). There is, however, one striking difference: rather than being stripped down, the astrolabe has a full rete with the remarkably large total of about 190 stars, many of which are labelled not just by name but also by planetary

FIGURE 4.3 Double spread of astrological diptych, by Thomas Hood, 1597 (British Library Board, Add. MS 71495).

temperament. The enhanced scope of the instrument is matched by its richer decoration. While some parts use the same blue, white and red colour scheme as the simpler version, the various degree and hour scales are coloured in dark blue, light blue, red and gold. In addition, the stars are picked out in gold and distinguished by symbols for the six standard magnitudes, conveniently summarized by a table inside the Tropic of Cancer on the mater. Unlike a conventional astrolabe, the rete sits above the limb rather than within it. With no physical constraint to terminate at the Tropic of Capricorn, two stars beyond the tropic are included: Markeb (k Pup) and Cor Scorpii (α Sco). The densely populated rete makes it impossible to include the table of planetary rulers which appears at the centre of the simpler diptych's astrolabe; this feature has been moved to the opposite sheet, freeing up space in the mater for a list of notable stars within the Tropic of Cancer. Distributed in two columns flanking the table of stellar magnitudes, the list is arranged alphabetically, with twenty-one stars (A–D) on the left and sixteen (E–S) on the right.

The remaining elements of this diptych correspond more or less exactly to the other's. Apart from the different content within the Tropic of Cancer, the mater of the more lavish astrolabe has the same layout and house texts as the simpler version. The opposite face is again in the format of a circular table and carries the same sections and scales present on the other example, but with the addition of the table of planetary rulers inserted before the volvelle of critical days at the centre. Accommodating more material required some compression: while the texts describing the signs are largely the same, some editing proved necessary. Thus Aries, with the longest entry on the simpler instrument, has here been abbreviated to fit the space available. The only other significant difference is that this diptych has no rotating index for the circular table. This, however, seems to be no more than an accident of survival; presumably one would originally have been provided for ease of identifying the many scales. Certainly, the layout of the volvelle of critical days could accommodate the same form of index as appears on the other version.[5]

What could an owner do with one of the diptychs? Knowing the Sun's place in the ecliptic, the astrolabes can display the ascendant and Medium Coeli for any particular time on that date. Then, with a set of ephemerides to deliver the positions of the planets, the houses which they occupy can quickly be read off and a figure erected for a nativity, election or interrogation. This work is latitude-specific: both the astrolabes are constructed for slightly more than 52°, appropriate to Cambridge, amongst other places.[6] The circular tables provide an aide-mémoire for subsequent detailed judgement, permitting a systematic reckoning of the dignities and debilities of the planets at a moment in time. They give in a single view a great mass of information that otherwise occupies many pages in textbooks. Finally, the full rete of Add. MS 71495 adds an extra layer of information, enabling the effect of powerful stars to be included in a more nuanced interpretation.[7]

Authorship and Audience

On the strapwork of the elaborate rete of Add. MS 71495 is the signature "Thomas Hood fecit 1597". Although the other diptych is not signed there can

5 When I first saw the two diptychs in 1997, they were incorrectly assembled. While Add. MS 71494 had its simple rete in place, its circular table carried the volvelle from Add. MS 71495 as well as its own volvelle and index ruler. Add. MS 71495 had a bare mater and its more elaborate rete was mounted over the circular table. The rearrangement to the current configuration was carried out by the British Library's conservation staff.

6 The latitudes are not explicitly stated; both have been derived from measurement of the distance from the north pole to the north point of the horizon.

7 The power of the fixed stars is discussed in Claudius Ptolemy, *Tetrabiblos*, ed. and trans. by Frank Egleston Robbins, Cambridge MA, Harvard University Press, 1940; reprint London, William Heinemann Ltd., 1980, (I, 9), pp. 46–59. For the appearance of planetary temperaments on astrolabes, see Silke Ackermann, "Astrological Scales on the National Maritime Museum Astrolabes", in Koenraad Van Cleempoel, *Astrolabes at Greenwich: A Catalogue of the Astrolabes in the National Maritime Museum, Greenwich*, Oxford, Oxford University Press and the National Maritime Museum, 2005, pp. 73–89: 81–82, 85–87.

be no doubt, based on content, format, handwriting and decoration, that Hood was also responsible for the simpler Add. MS 71494. The survival of the two diptychs adds to our knowledge of the range of Hood's work: astrology has not previously been identified as one of his active concerns. In addition, the more elaborate diptych was made at a very particular moment in Hood's career.

Hood studied at Cambridge, where he was a fellow at Trinity College and university mathematical lecturer for 1581–1582 before receiving a medical licence in 1584.[8] He next appears in 1588 as London's first public mathematical lecturer, a position created in the wake of the Armada crisis. Hood held this position for four busy years, actively publishing to complement his public teaching, devising new instruments for navigation and surveying, making sea charts and providing private tuition. When the lectureship was not extended he briefly continued his London mathematical work before apparently deciding that medicine would provide a more secure career. Although the next couple of years are undocumented, Hood seems to have returned to Cambridge and when he reappeared in 1595 he was now Doctor in Physic and seeking a licence from the Royal College of Physicians to practise medicine in London. If he had assumed his doctorate would ease him into the London medical elite, he was to be badly disappointed. Dismissed by the College's committee on 17 October 1595 for inadequate knowledge of Galen, it would take two more years and a further rebuff before he finally received conditional authorization from the College. During this awkward interim period before he could legitimately act as a physician, Hood returned to his former role, "knowing mine head and hands (under God) to be my best friends, I set them to their old occupation again, teaching the mathematical arts, and penning or translating such books as I thought most convenient for that purpose".[9]

The elaborate diptych therefore dates from the period when Hood again supported himself through private tuition in mathematics and advertised his London presence through publication. While the exact context for the simpler

8 The biographical details for Hood presented here come from Stephen Johnston, "Mathematical Practitioners and Instruments in Elizabethan England", *Annals of Science* 48 (1991), pp. 319–344; see also H.K. Higton, "Hood, Thomas (*bap.* 1556, *d.* 1620)", *Oxford Dictionary of National Biography*, Oxford, Oxford University Press, 2004, www.oxforddnb.com/view/article/13680 (accessed 25 Aug 2017).

9 From Hood's unpaginated dedication to his translation of Christian Urstitius, *The Elements of Arithmeticke*, London, Richard Field, 1596. Once licensed by the College of Physicians, the record of Hood's mathematical activity dries up. Ironically, despite his extended struggle to establish himself as a London physician, he did not stay in the capital. At an unknown date he set up in Gloucester and was still active there when he wrote his will in 1620. Outside the capital his MD and London licence were superfluous, as his original Cambridge medical licence from 1584 sufficiently qualified him for provincial practice.

of the two diptychs is unknown, their common features strongly imply that they would have been commissioned by similar clients. The exclusive use of Latin, the sheer labour involved in their intricate construction as well as the aesthetic commitment implied by their design and decoration, all indicate status-conscious production for the upper end of the market.

Hood had certainly aimed at high patronage before. An engraved pair of celestial planispheres he issued in 1590 was dedicated to Lord Lumley, who assembled one of Elizabethan England's most significant libraries.[10] These planispheres make a suggestive comparison with the diptychs, both because their stellar content overlaps with the elaborate rete of Add. MS. 71495 and because they are in Latin, even though (like Hood's other published works) the accompanying book was in English.[11]

A more concrete glimpse of the social world for which the diptychs were probably intended is afforded by Hood's only known pupil in the period from 1595 to 1597, when he had once again returned to the practice of mathematics. In December 1595, Arthur Throckmorton, brother-in-law to Walter Ralegh and a man of some substance familiar at court, began taking lessons in geometry from Hood. Apart from receiving the substantial sum of 20 shillings a week for his trouble, Hood immediately supplied Throckmorton with a range of drawing compasses. He also took a deposit for a sphere worth £10, indicating that he was expecting to pass on from geometry to astronomy and cosmography.[12]

Hood would surely have been no more than an intermediary for the mathematical instruments he supplied to Throckmorton. The expensive sphere—presumably an armillary sphere in brass—could have come from one of only a very few specialist makers in London. Hood had the personal connections to act as an agent, having worked with the engraver and instrument maker Augustine Ryther in the early 1590s and his apprentice Charles Whitwell later in the decade.[13] However Hood had no need to look elsewhere to source the astrological diptychs: he had the skills to make them himself.

---

10 Copies of Hood's *The Making and Use of the Geometrical Instrument Called a Sector*, London, Samuel Shorter, 1598, have bound into them a 1591 copperplate engraving of a cylinder dial by Leonaart Terwoort, again dedicated to Lumley in Latin by Hood.

11 Thomas Hood, *The Use of the Celestial Globe in Plano*, London, Tobie Cooke, 1590. A coloured pair of the planispheres is in the British Library, shelfmark Maps 184.h.1.

12 Johnston, op. cit. (n. 8), p. 336. On Throckmorton, see A.L. Rowse, *Ralegh and the Throckmortons*, London, Macmillan, 1962.

13 Ryther (*c.* 1550–1593) was responsible for the engraving of Hood's printed planispheres (1590) as well as a sea chart in 1592. Whitwell was named as the maker of the instrument described in Hood's 1598 book on the sector. On both, see Gerard L'E. Turner, *Elizabethan Instrument Makers: The Origins of the London Trade in Precision Instrument Making*, Oxford, Oxford University Press, 2000, pp. 27–31.

Though Hood has principally been studied through his published works, one strand of his mathematical activity was the commercial production of manuscript sea charts. At the time, this was a novel trade in England, and for historians of cartography Hood is a significant figure: the earliest known maker of manuscript sea charts in a London tradition whose style would continue for the next century.[14] Several signed and attributed examples are known and they display the same virtues of precise execution, neat penmanship and restrained decoration as the diptychs. What is especially striking is that navigation and astrology share the same innovative physical format. Hood's manuscript sea charts are on parchment, the earliest dated 1592. Such parchment charts had traditionally been rolled or stored flat. However, an example of Hood's work from 1596 is mounted on two wooden boards, the first example of what would become a distinctive feature of the so-called Thames School of chart-makers.[15] That the elaborate diptych uses this format in the very next year indicates just how closely interrelated were these aspects of Hood's work. Indeed, if the simpler, undated astrological diptych had been created significantly earlier than 1597, it is possible that this format—usually seen as peculiar to navigation—actually originated with astrology.

Mounting parchment on boards was of course intended to provide protection for the manuscript contents. There is some indication that the diptychs were used. They are not in pristine condition: from a distance they suggest the grubby appearance of an eventful life. It is only on closer inspection that the extraordinary care with which they were constructed really shines. More specifically, the four corners of each open diptych are damaged, with repairs to the underlying oak boards. Where the damage is least, the parchment and board are both visibly pierced. The holes, and the presence of traces of rust, suggest that iron pins may have been driven through the corners of each double spread, perhaps to mount the diptychs on a wall.

This is as far as we can currently go in determining the original audience and use of the diptychs. Unfortunately, their provenance does not supplement what can be gleaned from the diptych's physical form and what is known of Hood's career: they have so far been traced no further back than the middle of the nineteenth century (see *Appendix*). Given that both devices do effectively

14 Thomas R. Smith, "Manuscript and Printed Sea Charts in Seventeenth-Century London: The Case of the Thames School", in Norman J.W. Thrower (ed.), *The Compleat Plattmaker: Essays on Chart, Map, and Globe Making in England in the Seventeenth and Eighteenth Centuries*, Berkeley, University of California Press, 1978, pp. 45–100.

15 National Maritime Museum, Greenwich, G224:1/2, showing the Bay of Biscay and the English Channel. Reproduced and discussed in D. Howse, M. Sanderson, *The Sea Chart*, Newton Abbott, David and Charles, 1973, pp. 46–47.

the same thing, it is surprising that they first appear in the historical record together, as a pair. It seems unlikely (if not altogether impossible) that a single client would have commissioned both diptychs, one more luxurious than the other. But if they were created for different customers at different dates it is at least intriguing that they had been brought together by the mid-nineteenth century.

Sources

What existing materials did Hood draw on in designing and making the diptychs? What traditions do the devices align with? Taking the circular tables first, it is clear that their information can be found in many of the astrological textbooks of the period—whether one of the several Latin versions of time-honoured medieval classics such as the *Introduction* of al-Qabisi (Alcabitius) or more recent compilations such as Johannes Schöner's *Opusculum astrologicum*.[16] Nevertheless, from the great mass of literature there are two sources in particular that provide a compelling match to the content and arrangement of Hood's tables. They are both in the form of printed devices, and were published in the mid-sixteenth century, one by Gerard Mercator and the other by Claude Dariot. As for the astrolabes, there is no shortage of precedents for both the simplified and the elaborate versions. However, the design of the elaborate rete shows that Hood adopted a very specific pattern characteristic of the Louvain school of Mercator and the renowned Arsenius workshop.

In order to identify the sources for Hood's circular tables with confidence, it helps to briefly review the variety of astrological devices in sixteenth-century Europe. Against the background of this diversity, Hood's specific debts stand out all the more clearly. Some surviving items are remote from Hood's design in both scope and subject; for example, a manuscript disc in the Dibner Library which serves for astrological chorography, listing cities under the influence of each zodiac sign.[17] Other objects feature only a single element or a very restricted subset of the material incorporated by Hood. The prolific Georg Hartmann

16 Charles Burnett, Keiji Yamamoto, Michio Yano, *Al-Qabīṣī (Alcabitius): The Introduction to Astrology. Editions of the Arabic and Latin Texts and an English Translation*, London, Warburg Institute, 2004. Schöner's work was published in Nuremberg in 1539 and made extensive use of tables, providing information in as compendious a form as the constraints of movable type allowed.

17 Tayra Lanuza-Navarro, "An Astrological Disc from the Sixteenth Century", *Culture and Cosmos* 9, 2 (2005), pp. 63–67. This disc is perhaps missing a central aspectarium.

made a single-sided brass disc displaying the rulers of the planetary hours,[18] as well as a printed *Directorium Medicorum* (1538) which includes critical days; the latter topic is covered on other standalone discs and also occasionally appears as a small component of astrolabes.[19] Other tabular presentations of astrological material could be much more copious. The folding aide-mémoire of Heinrich Rantzau (dated 1574) consists of a pair of brass tablets packed with engraved and punched tables.[20] Much closer in time, place and personal connection to Hood is a 1596 gilt-brass disc attributed to the London instrument maker and engraver Charles Whitwell. Whitwell, with whom Hood was to collaborate in 1598 and who he surely knew before, used minuscule engraving to cram both historical chronology and astrology onto the two sides of a disc only 67 mm in diameter.[21] The skill needed for such fine work would have made this an expensive item, despite its size. It offers further, exactly contemporary evidence for the existence of an English audience with the wealth and enthusiasm to acquire rare and ingenious astrological devices of the kind represented by Hood's diptychs.

Closer in scale and form to Hood's circular tables is another print by Georg Hartmann (see *Fig. 3.4* in the chapter by Josefina Rodríguez-Arribas). It again bears the title *Directorium Medicorum* but on this occasion its various scales

18 The Hartmann brass disc of planetary rulers (1541) is now in the Museo Nacional de Ciencia y Tecnología, Madrid, inv. 1999/025/0003; see Marcos Villaverde Aparicio, “Astrología y religion: Georgius Hartmann en el Museo Nacional de Ciencia y Tecnología” in Josep Batlló Ortiz, Jordi Ferran Boleda, Mercè Piqueras Carrasco (eds.), *Actes de les VIII Trobades d'Història de la Ciència I de la Tècnica*, Barcelona, Societat Catalana d'Història de la Ciència i la Tècnica, 2006, pp. 299–306. The disc was previously in the Kunsthistorisches Museum, Vienna; see Hans G. Klemm, *Georg Hartmann aus Eggolsheim (1489–1586): Ein Mathematiker, Physiker und Instrumentenmacher von Europäischem Rang*, Eggolsheim, Markt Eggolsheim, 1989, p. 45.

19 Georg Hartmann's *Directorium Medicorum* includes a zodiac-calendar, lunar scales, an aspectarium and a set of instructions as well as a ring of critical days; Wolfenbüttel, Herzog August Bibliothek, shelfmark 15 Astron. 2° (1a), http://diglib.hab.de?grafik=15-astron-2f-00002 (accessed 6 April 2017). There is a disc of critical days by Erasmus Habermel at the Germanisches Nationalmuseum, Nuremberg; see Wolfgang Eckhardt, “Studien zu Erasmus Habermel: Erasmus und Josua Habermel—Kunstgeschichtliche Anmerkungen zu den Werken der Beiden Instrumentenmacher”, *Jahrbuch der Hamburger Kunstsammlungen* 22 (1977), pp. 13–74: 24, with several other versions and doubtful copies listed at pp. 69–70. For critical days on an astrolabe, see the combined astrolabe-equatorium at the Museum of the History of Science, Oxford, inv. 49847. See also the chapter by Rodríguez-Arribas in this volume, *infra*, pp. 51–59.

20 Günther Oestmann, *Heinrich Rantzau und die Astrologie: ein Beitrag zur Kulturgeschichte des 16. Jahrhunderts*, Braunschweig, Braunschweigisches Landesmuseum, 2004, pp. 75–85. See also the chapter by Oestmann in this volume, *infra*, pp. 92–103.

21 For Whitwell's disc, see Turner, op. cit. (n. 13), pp. 191–194.

overlap more extensively with Hood's content. Around a central aspectarium is a circular table which includes the signs of the zodiac, triplicities, a band combining detriment, exaltation and depression, the parts of the body ruled by the signs, the lunar mansions with their qualities, the lunar month and the qualities of its quarters, and a ring for critical days around the aspectarium.[22]

Although now fabulously rare, it is not inconceivable that Hood was aware of Hartmann's disc. But a much closer match comes from the printed astrological disc of Gerard Mercator, issued in 1551 to accompany his celebrated celestial globe of the same year (*Fig. 4.4*). Mercator provided a circular table with multiple concentric scales over which move indices for the seven planets and the head and tail of the dragon; these nine rules are free to rotate around a central aspectarium. Abbreviations for pitted, increasing fortune, lame, depression and rejoicing appear in the outer bands with the symbol and degrees of the zodiac signs. Lucid, dark, empty and smoky degrees are jointly placed in their own scale, as are masculine and feminine degrees, followed by decans, terms, triplicities, exaltations, and mansion and detriment. Although the content is mostly the same, the arrangement of Hood's material does not exactly match Mercator's, and the latter has some additional scales which Hood did not use, such as the terms according to Cardano and lunar mansions.[23]

Despite their differences of detail, the overall similarity between the comprehensive circular tables of Mercator and Hood is striking, as is the ambition to render the astrological division of the zodiac visible in a single view. Moreover, unless he had found a common source, Hood was copying Mercator's text. Closest to the centre of the Mercator disc are twelve texts describing the qualities of the zodiac signs. Hood absorbs the major part of these into his own longer texts for the signs, introducing them with additional material on the parts of the body over which each sign presides. Here again though, Hood was not slavishly following Mercator. Whereas Mercator included the parts of the body especially affected by each planet when in a particular sign, Hood presents this information in tabular rather than textual form, and his associations between the planets and individual body parts do not always correspond.

22 Germanisches Nationalmuseum, Nuremberg, HB 2713, dated to about 1554; a second copy is in Munich, Bayerische Staatsbibliothek, shelfmark Rar. 434, fol. 102; http://daten.digitale-sammlungen.de/bsb00028171/image_202 (accessed 6 April 2018).

23 Steven Vanden Broecke, "Dee, Mercator, and Louvain Instrument Making: An Undescribed Astrological Disc by Gerard Mercator (1551)", *Annals of Science* 51 (2001), pp. 219–240. Vanden Broecke provides a careful description and identifies the sources and context of Mercator's disc. Like Hartmann's astrological prints, Mercator's disc survives in a single copy, at the Historisches Museum, Basel, 1876.2.

FIGURE 4.4 Printed astrological disc, by Gerard Mercator, 1551 (Historisches Museum Basel, inv. 1876.20; Photo: P. Portner).

Mercator is persuasive but not precise as an inspiration for Hood's work. An even closer match for a large part of the content of Hood's circular tables is found in the work of the physician and mathematician Claude Dariot, whose *Ad astrorum iudicia facilis introductio* appeared in Lyon in 1557. A straightforward introductory work which also includes a tract on elections and material on critical days, it had a French translation the following year and was sufficiently marketable for an English version to appear in about 1583 and to be reissued as late as 1653. In the early part of the work, without any special fanfare, Dariot offered two woodcut tables to present his textual material in visual and interactive form (*Fig. 4.5* and *Fig. 4.6*). Each consists of a circular table with

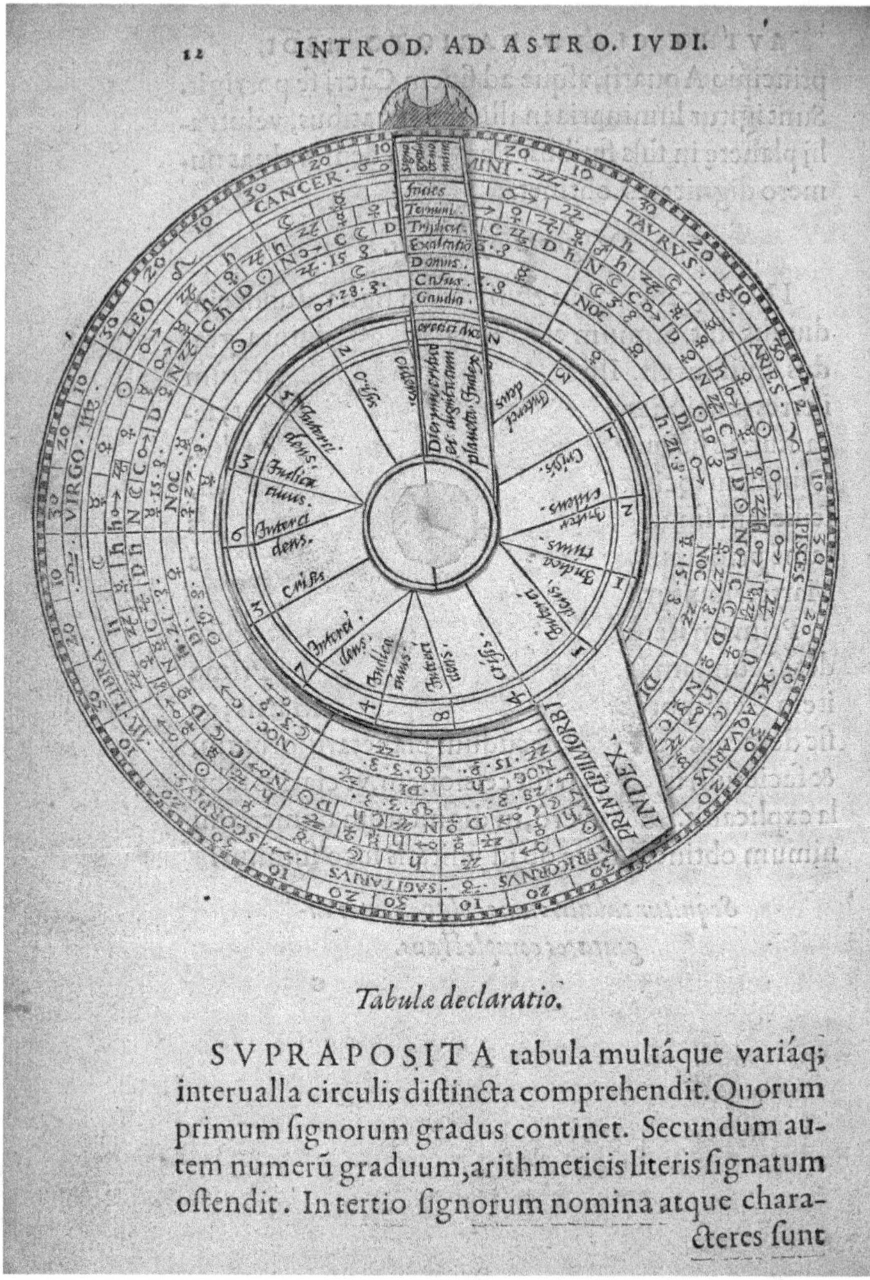
12 INTROD. AD ASTRO. IVDI.

Tabulæ declaratio.

SVPRAPOSITA tabula multáque variáq; interualla circulis diſtincta comprehendit. Quorum primum ſignorum gradus continet. Secundum autem numerũ graduum, arithmeticis literis ſignatum oſtendit. In tertio ſignorum nomina atque characteres ſunt

FIGURE 4.5 Circular table with volvelle and index, from Claude Dariot, *Ad astrorum iudicia facilis introductio*, Lyon, Apud Mauritium Roy et Ludovicum Pesnot, 1557, p. 12 (British Library Board, shelfmark 718.e.45).

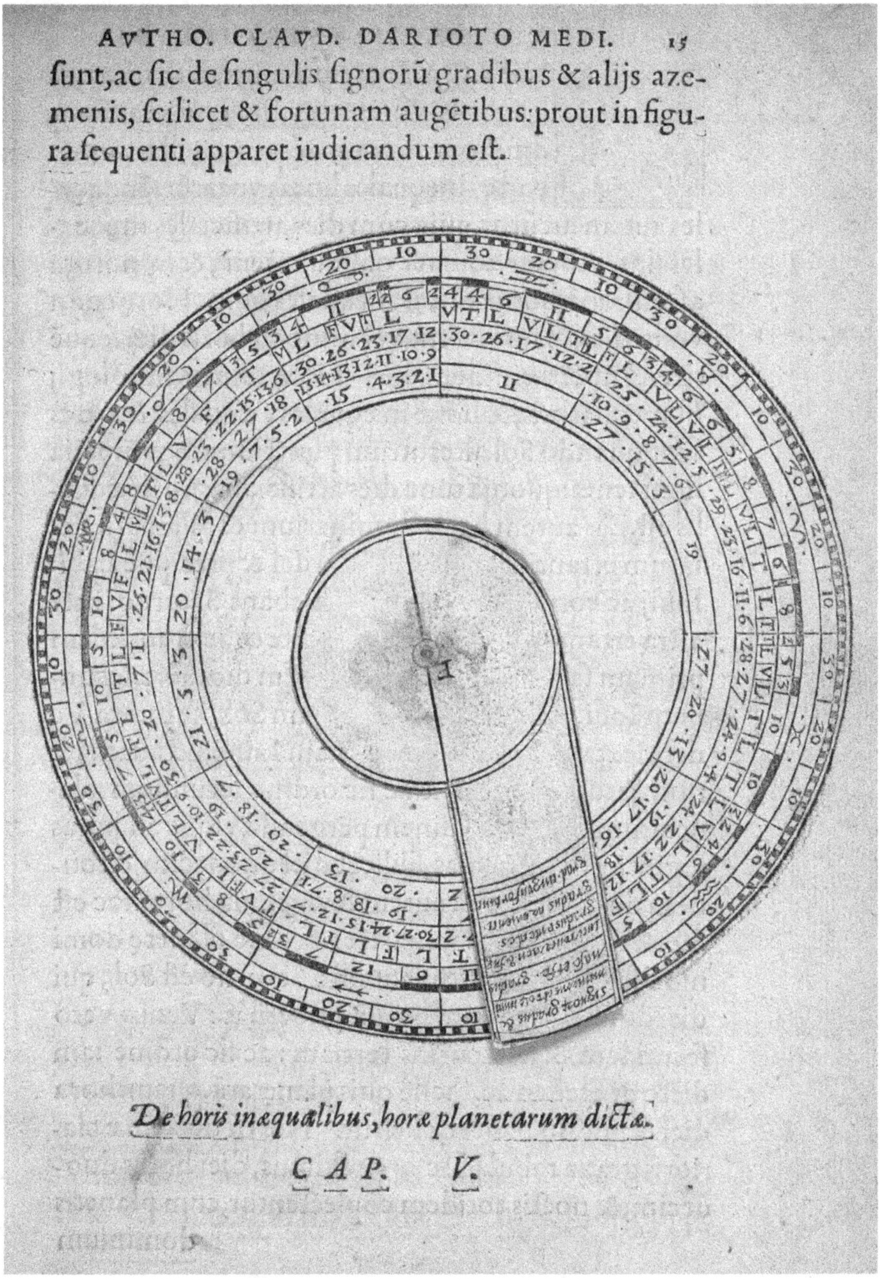

AVTHO. CLAVD. DARIOTO MEDI. 15

ſunt,ac ſic de ſingulis ſignorũ gradibus & alijs azemenis, ſcilicet & fortunam augẽtibus: prout in figura ſequenti apparet iudicandum eſt.

De horis inæqualibus, horæ planetarum dictæ.

CAP. V.

FIGURE 4.6 Circular table with index from Claude Dariot, *Ad astrorum iudicia facilis introductio*, Lyon, Apud Mauritium Roy et Ludovicum Pesnot, 1557, p. 15 (British Library Board, shelfmark 718.e.45).

a rotating index identifying the underlying scales; the first also has a rotating volvelle for critical days at its centre.

Although split across two devices, the correspondence in format, content and sequence with Hood's tables is remarkable. Dariot uses the same twelve scales to divide the zodiac, with the same terminology. After the zodiac signs around the circumference, his first table carries: faces; terms; triplicities; exaltations; mansions; depression and detriment; and rejoicing. The second table's outer ring of zodiac signs surrounds: masculine and feminine degrees; light, dark, smoky and void degrees; pitted; lame; and augmenting fortune. This is even the same order as Hood would use when combining the two tables into one, with a single alteration: Hood swaps the position of two contiguous scales, placing rejoicing before the scale for depression and detriment. Although Hood did not mimic every element of the woodcut designs (for example, he did not follow the marking of masculine degrees as black and feminine as white) there can be little doubt that he used Dariot's presentation in devising the set of scales and the volvelle of critical days in his own circular tables. The conclusion is underlined by another circular table which Dariot places shortly after the first two. This is for the planetary hours, with a rotating index of days, which Hood would also incorporate in his diptychs.[24] In devising his circular tables, Hood therefore seems to have directly borrowed from Dariot, presumably in the Latin, with Mercator acting as both a specific textual source and the more general inspiration towards a single device encompassing as much astrological material as possible.[25]

---

24 The table of planetary hours appears in the Latin edition: Claude Dariot, *Ad astrorum iudicia facilis introductio*, Lyon, Apud Mauritium Roy et Ludovicum Pesnot, 1557, p. 17. The table also appears in the French version (Dariot, *L'introduction au jugement des astres*, Lyon, Par Maurice Roy et Loys Pesnot, 1558, p. 26) but was omitted from the English translation, suggesting that Hood did not use it. Not all owners went to the trouble of cutting out and attaching the indices and volvelles in Dariot's work. Several copies are bound with the uncut sheet carrying the book's four moving parts. Particularly interesting is the copy in the Folger Shakespeare Library (callmark 164- 971q), whose large uncut sheet contains not just the moving parts but also the printer's instructions on how to make them up. See the blog post http://collation.folger.edu/2014/09/constructing-volvelles/ (accessed 6 April 2018).

25 No source has yet been traced for the other major textual component of Hood's diptychs, the descriptions of the houses on the maters. Mercator would have been of no help because his synopsis of the houses appeared not on the astrological disc but on the horizon ring of the 1551 celestial globe. The limited space there meant each house could be spared only about twenty words, much shorter than Hood's accounts.

What of Hood's overall format, combining astrolabe and circular table into a diptych? Mercator certainly made astrolabes, but his disc was meant to accompany a celestial globe.

An astrolabe and circular astrological table do feature together in at least one sixteenth-century manuscript, though not as a diptych: the volume contains several other pasteboard instruments.[26] There is no reason to think that Hood was familiar with this specific volume, which is unlikely to be English and whose devices have a different design and layout. But it shows that others had already been trying similar juxtapositions.

The most tantalizing point of comparison for Hood's diptych format is an incomplete astrolabe at Greenwich. Only the mater survives, perhaps because it was reused in the seventeenth century, but the back copies the scales on Mercator's astrological disc. It has been attributed to the London-based Flemish immigrant Thomas Gemini, who was closely linked with Mercator and began the commercial manufacture of mathematical instruments in England. The astrolabe has been dated to immediately after Mercator's publication, since it carries the arms of Edward VI (reigned 1547–1553).[27]

It is tempting to think that Hood could have seen this high-status astrolabe with its circular astrological table on the reverse, and devised his diptychs in response. But there are reasons to think that he did not. To reach that conclusion we need to turn away from the circular tables and their sources to investigate the design of the astrolabes in Hood's diptychs.

Other than texts, astrolabes were the most common Renaissance astrological object. While some examples had unusual and special-purpose astrological features, almost all carried the planetary hours and house lines on their latitude plates. These features are so commonplace that historians and curators are now apt to treat them as merely conventional presences, or overlook them altogether. But they doubtless had a more tangible presence for many Renaissance owners.

Like Hood's simple astrolabe, a few Renaissance instruments in brass have similarly stripped-down retes—the absence of stars suggesting that they were

---

26 Oxford, Museum of the History of Science, MS Radcliffe 74. For a summary listing of the manuscript's contents, see John North, *Richard of Wallingford: An Edition of his Writings*, Oxford, Clarendon Press, 1976, 3 vols.; v. 3, pp. 222–223.

27 National Maritime Museum, Greenwich, AST0567; see Van Cleempoel, op. cit. (n. 7), pp. 146–152; Gerard L'E. Turner, Koenraad Van Cleempoel, "A Tudor Astrolabe by Thomas Gemini and Its Relationship to an Astrological Disc by Gerard Mercator of 1551", *The Antiquaries Journal* 81 (2001), pp. 400–409. A transcription of the astrological scales is given in Ackermann, op. cit. (n. 7), pp. 82–84. Gemini's career and relationship with Mercator are discussed in Turner, op. cit. (n. 13), pp. 12–20.

principally intended for astrological use, especially when the latitude plates dispense with the usual grid of almucantars and azimuths.[28] Hood's simple rete suggests familiarity with this existing aspect of the repertoire of Renaissance astrolabe making.

However, it is the elaborate rete that provides a more fruitful avenue for investigation. The key feature of this rete is that it is of the tulip type associated with Louvain astrolabe makers. These retes are formed of three intertwining straps which provide the framework to support the star pointers. The characteristic tulip shape is inside the ecliptic circle. The design became something of an international brand in the sixteenth century through the wide distribution of the astrolabes of Mercator and the Arsenius workshop, as well as their successors such as Adrian Descrolières. It was carried to London by Thomas Gemini and also adopted in Prague by Erasmus Habermel.[29]

Thanks to the studies of Gerard Turner and Koenraad Van Cleempoel it is possible to reach surprisingly specific conclusions about the exact pattern that Hood followed, and therefore the instruments that he had most likely seen. The basic tulip pattern is easy enough to identify, picked out by Hood in red. Indeed, more clearly than in brass, Hood reveals that each of the three straps forms a distinct closed loop. While he continues the *trompe l'oeil* intertwining of the brass instruments' strapwork, he enhances it by outlining each strap with its own colour (blue, off-white or red).

But it is the detailed shape of the straps' tracery that is particularly revealing. As Van Cleempoel has documented for Mercator and the Louvain instrument makers, the tulip design was continuously developed from the 1540s to the end of the century, with each instrument carrying its own subtly different signature form. Although none is quite identical, there are nevertheless key features that allow a typology to be constructed (*Fig. 4.7*). Thus Mercator used an early form in the 1540s, which Gemini adopted and adapted in his two surviving retes from the 1550s (type 1 in Van Cleempoel's typology). In later work

28 For example, an astrolabe by Johannes Wagner, Nuremberg, 1538, at the Museum of the History of Science, Oxford, inv. 40443, which has a rete with only the ecliptic (each sign having the symbol of its planetary ruler) and two plates which contain only the houses and planetary hours. A larger example, perhaps from the late-fifteenth century, at the Pinacoteca Ambrosiana in Milan is reproduced as the frontispiece to John North, *Horoscopes and History*, London, Warburg Institute, 1986. Both these are referred to by Oestmann, *Heinrich Rantzau*, cit. (n. 21), pp. 67–72, along with four other examples in Nuremberg, Cambridge, Munich and London.

29 For Habermel, note the large astrolabe with tulip rete at the Museum of the History of Science, Oxford, inv. 38097, with a table of the planetary rulers of the hours on the reverse. This was a deliberate choice by Habermel: he made other astrolabes with distinctively different rete designs, such as inv. 37297 in the same collection.

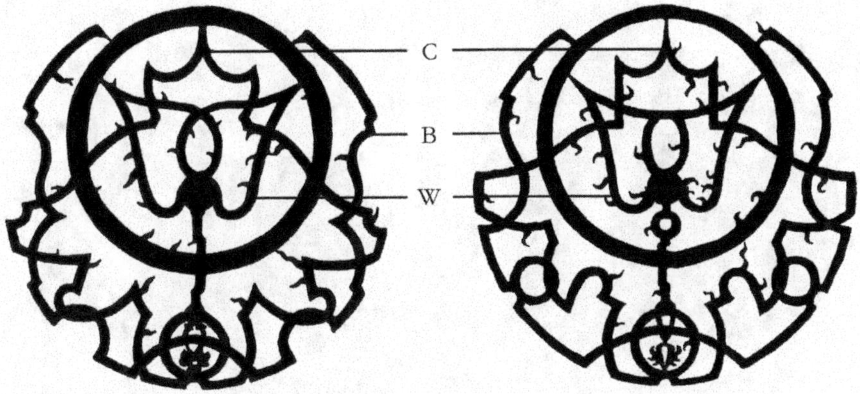

FIGURE 4.7 Tulip rete silhouettes of Gerard Mercator (type 1, left) and Gualterus Arsenius (type 2, right), adapted with permission from Koenraad Van Cleempoel, *A Catalogue Raisonné of Scientific Instruments from the Louvain School: 1530 to 1600*, Turnhout, Brepols, 2002.

of around 1570, Mercator adopted a pattern developed by Gualterus Arsenius in astrolabes dated between 1554 and 1557 (type 2). Arsenius and his successors would go on to introduce further artistic novelties, varying the width of the straps and devising new motifs to create metalworking *tours-de-force*.[30]

Hood's design is unique but most closely related to Van Cleempoel's type 2. To pick out its distinguishing features, it helps to be able to differentiate the three straps. Van Cleempoel refers to the tulip-shaped strap within the ecliptic circle as W. The second strap B is exclusively outside the ecliptic and the longest strap C winds across the whole of the rete. At the bottom of the rete, the B strap of both type 1 and type 2 retes makes a bridge passing through a central circle formed by strap C. It is at just this significant spot that Hood placed his signature. One of the key discriminants between type 1 and type 2 is the presence in type 2 of two additional circles flanking the central one. They are formed by strap B, with strap C passing through. Hood has created a close variant of these circles, not documented on any surviving metal astrolabe. The circles themselves are raised above the Tropic of Capricorn, much more prominently than in the type 2 astrolabes of Arsenius in the 1550s and of Mercator c. 1570. Moreover, rather than simply passing through the circles, Hood's strap C impales them (*Fig. 4.8*).

30 My discussion here derives from Koenraad Van Cleempoel, *A Catalogue Raisonné of Scientific Instruments from the Louvain School: 1530 to 1600*, Turnhout, Brepols, 2002, particularly pp. 46–50.

FIGURE 4.8 Detail of the rete of Thomas Hood's tulip astrolabe showing his signature inside the central strapwork circle as well as the two impaled flanking circles (British Library Board, Add. MS 71495).

In terms of design, Hood appears to have made a modest original contribution to the astrolabe maker's art, or copied a now-lost instrument. Whether he was creatively adapting or directly copying, he must have had access to at least one of the highest quality instruments of the period. The only makers who created type 2 retes in the form which inspired Hood were Arsenius and Mercator. After 1557, Arsenius stopped using the two additional circles and the last instruments to feature them are the two Mercator astrolabes of *c.* 1570 now in Augsburg and Florence.[31]

One implication of this design review is that—tantalizing as it seemed earlier—Hood was probably not following the incomplete astrolabe by Thomas Gemini that carries Mercator's astrological scales. The absence of the rete from Gemini's instrument is frustrating, as it prevents a conclusive comparison with Hood. But a strong inference can nevertheless be drawn. Gemini's two surviving retes are from astrolabes of the late 1550s, made some years after the incomplete example now in Greenwich which dates to no later than 1553. Both of these later retes follow the type 1 pattern of Mercator's first surviving astrolabe of *c.* 1545. It therefore seems all but inconceivable that Gemini's earlier Greenwich instrument could have included a rete of the type 2 pattern, introduced in about 1554 by Arsenius.

31 For the Mercator astrolabes, see Gerard L'E. Turner, "The Three Astrolabes of Gerard Mercator", *Annals of Science* 51 (1994), pp. 329–353.

We will most likely never know what specific instruments Hood had seen before designing his elaborate rete. But it is possible to say with some confidence that he was following the patterns established by Mercator and Gualterus Arsenius. A product of the Arsenius workshop, specifically from the mid-1550s, would have provided Hood with the ideal starting point for his elaborate rete. Not only do those astrolabes provide the closest match for Hood's design but, between Mercator, Gemini and Arsenius, it is only the latter who added planetary temperaments to the star names on his retes, just as Hood would later do.[32] Hood thus provides new evidence for the international transmission and reception of high-end mathematical instruments.

Conclusion

This study has had a deliberately limited focus, seeking to establish specific conclusions on the purpose, audience and sources of Thomas Hood's diptychs, and doing so directly from the evidence of the objects themselves and the immediate circumstances of Hood's career.

One difficulty with Hood's instruments is that they do not come with instructions or a programmatic statement that would supply a larger sense of intellectual context. Gerard Mercator's printed astrological disc appeared with an address to the reader on its reverse, which Steven Vanden Broecke has used to detect Mercator's reforming ambitions for astrology and his device's astronomical and philosophical setting. Anything Hood may have supplied does not survive.

This is not to say that Hood lacked an intellectual framework within which to situate the diptychs. On the contrary, several of his published writings refer to astrological matters, and suggest that Hood's diptychs can best be understood in relation to the tradition of cosmographic astrology represented by the work of Mercator and his Louvain teacher Gemma Frisius.[33] To demonstrate that claim would require a fuller study than can be accommodated here, drawing particularly on Hood's *Use of the Celestial Globe in Plano* (1590) and *The Use of Both the Globes, Celestiall and Terrestriall* (1592). Just as Mercator's astrological disc complemented his celestial globe of 1551, so Hood's diptychs can be juxtaposed with these writings on his own engraved celestial planispheres

32 In Van Cleempoel, op. cit. (n. 30), only the earliest surviving Arsenius astrolabe (now in Mexico City) does not carry planetary temperaments (p. 87ff).

33 Steven Vanden Broecke, *The Limits of Influence: Pico, Louvain, and the Crisis of Renaissance Astrology*, Leiden, Brill, 2003, pp. 113–136.

and the use of the new English globes of Emery Molyneux. The present study serves as a stepping stone towards that broader ambition.

In his text on Molyneux's globes Hood commented that "we take this for a generall rule in any Mathematicall instrument, that the greater it is the more precisenesse and pleasure is in the use thereof."[34] The same values evidently underpinned the preparation of his diptychs. They were large, neatly constructed and clearly intended as objects of both intellectual and visual consumption. Combining tabular astrological doctrine with the familiar form of the astrolabe, they delivered knowledge in compendious form and supported by instrumental action. The diptychs straddle conventional boundaries between both text and instrument and practical problem-solving and conspicuous display. They would have appealed to individuals with a taste for mathematical neatness and elegance and both the education and appetite for Latin learning. While the diptychs could be used didactically to present the science of the stars in decorative form, their undoubtedly high cost restricted them to a necessarily limited audience. By joining astrological orthodoxy to artful design, Thomas Hood created objects to rival the finest products of contemporary cosmography.

Appendix

This appendix traces the provenance of Hood's astrological diptychs, for two reasons. One is to encourage further research on their provenance, in the hope that it can be pushed back beyond the mid-nineteenth century. Documenting the currently known sources might enable others to recognize or identify earlier records. The second reason is to appreciate how the diptychs came to survive at all. The stereotyped image of manuscripts slumbering in an archive until the revelation of their scholarly "discovery" typically hides the actions and efforts of a long series of dealers, collectors and librarians. In this case, it also obscures the surprising way in which astrology was recruited into the politics of Victorian knowledge.

The provenance of the diptychs can most readily be revealed by working backwards. They were accessioned by the British Library Manuscripts Department in 1994. They came to Bloomsbury from the Patent Office Library, which from 1962 had been administered by the British Museum's library department as the National Reference Library of Science and Invention. The diptychs

34 Thomas Hood, *The Use of Both the Globes, Celestiall and Terrestriall*, London, Thomas Dawson, 1592, p. D[1]r.

carry multiple stamps of the Patent Office Library as well as marks of its cataloguing. They are not listed in the Library's original published catalogue of 1857–1858 but appear in both the 1881–1883 and 1898 versions of the catalogue. In the latter they are listed as "Hood, Thos., M.D., Lecturer on Mathematics. [Astrological Diagrams. MS. on vellum, enclosed in oaken cases. Circa 1597.]".[35]

Remarkably, what appears to be the original Patent Office Library accession slip for the diptychs survives in the British Library. Aside from the same description as would later appear in print it carries an additional note: "(From Maynard, & attributed by him to Thos. Hood, D.M. Mathematical Lecturer in the City of London temp. Eliz.)".

This provides a vital clue to the diptychs' previous history. Maynard is Samuel Maynard, listed in the 1852 Post Office London Directory (Small Edition) as "Maynard Samuel, mathematical & philosophical bookseller, 8 Earl's Court, Cranbourn Street, Leicester Square" (p. 878). Maynard was a notable and somewhat idiosyncratic specialist bookseller for some 30 years. His stock can be traced thanks to the bibliophile mathematician Augustus De Morgan, who collected the catalogues of booksellers and regularly bound them together. These volumes are now housed with the rest of his library in the University of London's Senate House Library.

Maynard's catalogue XI is included in a volume assembled by De Morgan in September 1854 (shelfmark B.P.359), and presumably dates from that year. Item 943 begins "HOOD (T.) Astrological Diagrams, on four Sheets of vellum, enclosed in Oak Cases, exhibiting at one view a complete System of Astrology in Theory and Practice". A description then follows in which the circular diagrams are characterized as the theory and the astrolabes are identified as "Two Diagrams of the twelve Houses", representing practice. Maynard devotes far more space to these evidently unusual items than to his standard printed stock. His description ends with the attribution to "Thos. Hood, 1597. Elaborately written, gilded, and coloured, by which every division of the Subject is rendered most beautifully clear and distinct".

However, this 1854 catalogue was not the source for the Patent Office's acquisition. Another of De Morgan's *sammelbands* (B.P.357) contains Maynard's fourteenth and last catalogue and also both a manuscript biographical note of Maynard and an informed and affectionate printed obituary (perhaps by De Morgan himself?). Bound immediately after Maynard's last catalogue is

35 *Catalogue of the Library of the Patent Office*, London, Her Majesty's Stationery Office, 1898, 2 vols.; v. 1, p. 417. The diptych's original catalogue reference of 6876 was subsequently changed under a new classification to FB 65/11 in the early twentieth century. FB 65 was the Patent Office's (much-populated?) section for general treatises on astrology.

Sotheby's auction catalogue of his stock, sold over three days in January 1863, after his retirement. Lot 1178—the last item on the last day of the sale—was Hood's astrological diptychs. The diptychs thus remained unsold from 1854 to 1862; the obituary clipped by De Morgan says "Mr Maynard carried on his business on the plan of high prices and slow returns: this is not the way to thrive in our day". Hood's instruments must therefore have been acquired by the Patent Office from the Sotheby's sale of 1863.

Why, in the middle of the nineteenth century, was the recently-founded Patent Office interested in sixteenth-century astrological instruments? Their acquisition can be seen as one of the more curious outcomes of a long and energetic campaign waged by Bennet Woodcroft, assistant to the commissioner of patents after the passing of the Patent Law Amendment Act in 1852. At the time, patents were controversial and far from universally approved. Patent abolitionists argued that invention was a social response to need, which unjustly rewarded only one among the many who might have a reasonable claim on any particular invention. Bennet Woodcroft sought to rebut their arguments. By building up the Patent Office Library and its associated Museum, Woodcroft not only made the history of invention more accessible for practical purposes of information retrieval but also to celebrate and reward the unjustly neglected figure of the inventor.[36] Hood's diptychs were probably seen as early evidence of English invention and their preservation a way of recognizing the work of an otherwise obscure figure.

The survival of the compendia is thus a result of the specific efforts of Bennet Woodcroft in support of the Patent Office. It is also an outcome of the more general contemporary interest in the history of science and mathematics, in which the historical researches of an Augustus De Morgan helped to create a market for antiquarian science served by such specialist booksellers as Samuel Maynard.

The identification of Maynard as the bookseller who acquired and offered the diptychs helps to explain some of their other features. The cover of Add. MS 71494 (the diptych with the simple astrolabe) has pasted on it Maynard's descriptive notice from his catalogue XI. This had been tightly trimmed so that its origins were not apparent. The diptychs have two other associated documents, both from the nineteenth century. Pasted below Maynard's printed notice is a manuscript description of similar scope and wording. Accompanying the diptychs is a longer manuscript sheet which opens with a description matching

36 Christine MacLeod, "Concepts of Invention and the Patent Controversy in Victorian Britain", in Robert Fox (ed.), *Technological Change: Methods and Themes in the History of Technology*, Amsterdam, Harwood Academic, 1996, pp. 137–154: 143.

the wording of the printed notice. This is followed by instructions for using the astrolabe to determine the houses, with two worked examples. However, it is evident that this is not a descendant of an original sixteenth-century set of instructions: the planetary hour lines under the horizon of the mater are confused with the cusps of the houses. This accompanying sheet is written on the reverse of a small section of a large-scale city map of London. On the basis of the buildings that appear (particularly the new Westminster Bridewell of 1831 and Buckingham Palace before the addition of the East Wing of 1847–1850 that enclosed the courtyard) this map can be dated to the 1830s or 1840s.

CHAPTER 5

Displaying Astrological Knowledge through Tabulation: Some Notes Pertaining to Particular Arrangement on Instruments

Günther Oestmann

Time measurement is an indispensable prerequisite for astrology, so any timepiece—be it a sundial, astrolabe or mechanical clock—can serve astrological purposes. Occasionally, explicitly astrological information, such as the tabulation of the lords of time (also called the chronocrators or planetary rulers), can be found on astrolabes and sundials.[1] The astronomical clock in Münster Cathedral (1540–1542) is equipped with mechanically operated tablets for the indication of the chronocrators,[2] while the clock in Rostock Cathedral has a

1 Astrological information on scientific instruments and clocks has been given occasional attention: Harrold E. Gillingham, "Astrological Instruments", *Antiques* 32 (1928), pp. 32–34; Ernst von Bassermann-Jordan, "Astrologie, Uhrmacher und Uhren", *Deutsche Uhrmacher-Zeitung* 54 (1930), pp. 794–798 and 856–857; Sharon L. Gibbs, "Astronomie und Astrologie—ihre Indikation an Uhren", in Klaus Maurice, Otto Mayr (eds), *Die Welt als Uhr: Deutsche Uhren und Automaten 1550–1650*, Munich, Bayerisches Nationalmuseum, 1980, pp. 55–62; Silke Ackermann, "Astrological Scales on the National Maritime Museum Astrolabes", in Koenraad Van Cleempoel, *Astrolabes at Greenwich: A Catalogue of the Astrolabes in the National Maritime Museum Greenwich*, Oxford, Oxford University Press, 2005, pp. 73–89; see also Elly Dekker, *Illustrating the Phaenomena: Celestial Cartography in Antiquity and the Middle Ages*, Oxford, Oxford University Press, 2013, pp. 388–408. Helge B.J. Karlsen, "Astrologie und Zeitmessung vor 400 Jahren: Das Zifferblatt für die Stundenregenten (Gubernatores) an einer großen Tischuhr von Johann Reinhold, Augsburg", *Schriften des historisch-wissenschaftlichen Fachkreises "Freunde alter Uhren" in der Deutschen Gesellschaft für Chronometrie* 21 (1982), pp. 39–45. On a sundial made in 1550 by Zacharias Scultetus (1530–1560) with planetary houses and chronocrators, see Walter Zimmermann, "Die Sonnenuhr an der Ratsapotheke in Görlitz", *Neues Lausitzisches Magazin: Zeitschrift der Oberlausitzischen Gesellschaft der Wissenschaften* 111 (1935), pp. 142–168, and on the origin and dissemination of the planetary week in antiquity see Francis Henry Colson, *The Week: An Essay on the Origin and Development of the Seven-Day Cycle*, Cambridge, Cambridge University Press, 1926, pp. 18–61; Franz Boll, "Hebdomas", in *Paulys Realencyclopädie der classischen Altertumswissenschaft*, Stuttgart, Metzler, 1893–1980, v. 7, cols. 2556–2578; Wilhelm and Hans Georg Gundel, "Planeten", in *ibid.*, v. 20, cols. 2143–2147.

2 Theodor Wieschebrink, *Die astronomische Uhr im Dom zu Münster*, Münster, Aschendorff, 1983, pp. 46–48; Claus Peter, "Zum astronomischen Zeigerwerk und einigen Sondereinrichtungen

 | DOI:10.1163/9789004381438_007

gravity-operated dial.[3] But these mechanisms indicate hours of equal length, which is in contradiction to the definition of astrological planetary hours (i.e. division of the diurnal and nocturnal arc of the Sun into twelve hours each, which are consequently of unequal length). This chapter looks at examples on some smaller instruments, which provide important data and interpretative tools for horoscopic work, and for which the term 'tabular instrument' may be coined. Examples will be discussed with special reference to a unique device formerly owned by Heinrich Rantzau (1526–1598), an ardent believer in astrology.

The Astrolabes of Tobias Volckmer

Tobias Volckmer the Elder (*c.* 1550–1622), a goldsmith and instrument maker from Braunschweig, had a particular liking for the display of the chronocrators, which figure prominently on his astrolabes (*Fig. 5.1*).[4] Volckmer was a versatile and skilled goldsmith, instrument maker, mathematician and surveyor who had been trained from 1565 onwards in the workshop of Jeronimus Remmerdes (Remmers, died 1602), a goldsmith in Braunschweig. He received citizenship in

---

der astronomischen Uhr im Dom zu Münster (Westfalen)", *Beiträge zur Astronomiegeschichte* 13, "Acta Historica Astronomiae 58", Leipzig, Akademische Verlagsanstalt, 2016, pp. 158–159.

3 Helge Tideke [identical with Helge B.J. Karlsen (n. 1)], "Die geniale Scheibe der Planetenstunden an der astronomischen Uhr in St. Marien zu Rostock", *Mitteilungen der Deutschen Gesellschaft für Chronometrie* 143 (2015), pp. 25–26; Manfred Schukowski, "Die zwei Scheiben auf dem Stundenzeiger der astronomischen Uhr in der Rostocker Marienkirche", *Beiträge zur Astronomiegeschichte* 13, "Acta Historica Astronomiae 58", Leipzig, Akademische Verlagsanstalt, 2016, pp. 132–136.

4 Although there are a number of relevant publications, Tobias Volckmer's life and work has not yet been fully researched. Gerd Spies, *Braunschweiger Goldschmiede: Geschichte, Werke, Meister und Marken*, München/Berlin, Klinkhardt & Biermann, 1996, 3 vols.; v. 3, pp. 78–79; Karl Gerke, "Tobias Volkmer aus Braunschweig", *Mitteilungen der Technischen Universität Carolo-Wilhelma zu Braunschweig* 9 (1974), pp. 32–38; Franz Wagner, "Zwei protestantische Künstler am Hof der Salzburger Erzbischöfe", in Friederike Zaisberger (ed.), *Reformation, Emigration: Protestanten in Salzburg*, Salzburg, Amt der Salzburger Landesregierung/ Kulturabteilung, 1981, pp. 69–70; Karl Gerke, "Tobias Volkmer", in Cord Meckseper (ed.), *Stadt im Wandel: Kunst und Kultur des Bürgertums in Norddeutschland*, Stuttgart/Bad Cannstatt, Edition Cantz, 1985, 4 vols.; v. 1, pp. 593–596; Dietrich Mack, *Braunschweiger Bürgergeschlechter im 16. und 17. Jahrhundert: Genealogien von Stiftern der 3 Bildzyklen in der Brüdern-Kirche zu Braunschweig (1596–1638)*, Göttingen, Goltze, 1985, 3 vols.; v. 2.3, pp. 835–838 (on Volckmer's genealogy); Max Frankenburger, *Die Alt-Münchner Goldschmiede und ihre Kunst*, München, Bruckmann, 1912, pp. 323–324 and 347–348; Siegmund Günther, "Die beiden Münchener Geometer und Kartographen Tobias Volckmar", *Jahrbuch für Münchener Geschichte* 5 (1894), pp. 1–16.

FIGURE 5.1 Astrolabe, by Tobias Volckmer the Elder, 1593; detail of back (Braunschweigisches Landesmuseum, Braunschweig, LMB 22994).

Salzburg in 1586 and worked as court goldsmith for three archbishops (Johann Jacob Khün von Belasy, Georg von Kuenburg and Wolf Dietrich von Raitenau) until 1593. Volckmer was Protestant and during the Counter-Reformation his position apparently became difficult. He tried in vain to secure a position in Dresden, but was successful in finding an appointment in Munich as court goldsmith and mathematician in 1594. He not only made instruments there, but was also active as a cartographer and surveyor. He took part in the restoration of the fortress of Ingolstadt in 1609, and with his son (of the same name, died 1658) he designed the layout of the "Hofgarten" in Munich as well as surveying the area.[5] Together they also surveyed the route for a saltwater pipeline from Reichenhall to Traunstein in 1616.[6] Three surviving astrolabes by Volckmer display a huge range of astrological data on the back. Two were made in Salzburg, one dated 1594 (*Fig. 5.2*), the other 1593 (*Fig. 5.3*); the third was made in Munich and dated 1601 (*Fig. 5.4*).[7]

5 Brigitte Volk-Knüttel, "Der Münchner Hofgarten und die Goldschmiede Tobias Volckmer Vater und Sohn", *Münchner Jahrbuch der bildenden Kunst* 64 (2013), pp. 69–94.

6 Heinrich Kurtz, "Die Soleleitung von Reichenhall nach Traunstein 1617–1619: Ein Beitrag zur Technikgeschichte Bayerns", *Abhandlungen und Berichte des Deutschen Museums München* 46 (1978), H. 1/2, Munich, Oldenbourg, p. 56.

7 British Museum, London, inv. no. 1888, 1201.280; Braunschweigisches Landesmuseum, Braunschweig, inv. no. LMB 22994; and Stift Seitenstetten, Austria. The late P. Benedikt

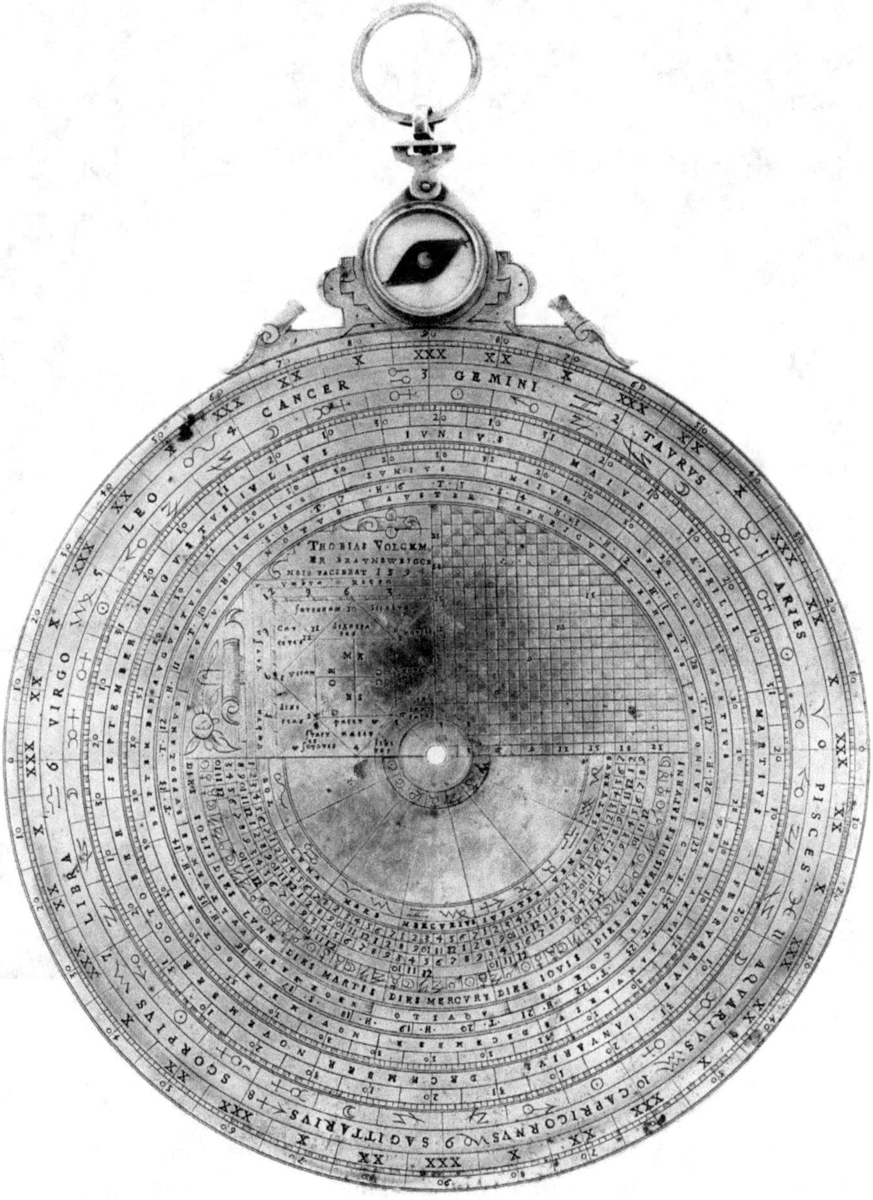

FIGURE 5.2 Astrolabe, by Tobias Volckmer the Elder, 1594; back (British Museum, London, 1888, 1201.280).

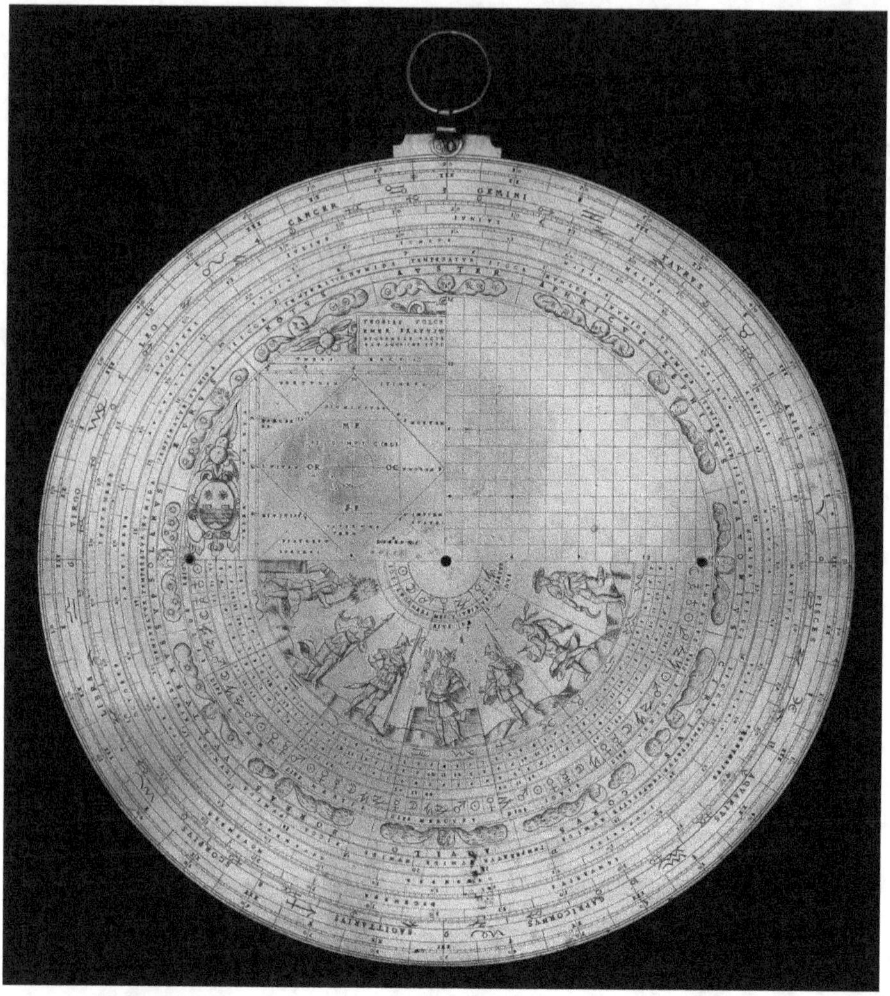

FIGURE 5.3 Astrolabe, by Tobias Volckmer the Elder, 1593; back (Braunschweigisches Landesmuseum, Braunschweig, LMB 22994).

In addition to astronomical and mathematical problems connected with astrology (e.g. house division), specific instruments could serve to display in a clear arrangement the entangled, diverse astrological patterns needed for an interpretation, including the meanings of the houses, properties of a planet, its position in the houses and signs of the zodiac, the properties of the zodiac

---

Wagner (1929–2012), formerly registrar of the Benedictine monastery at Seitenstetten, kindly provided photographs of the instrument held there.

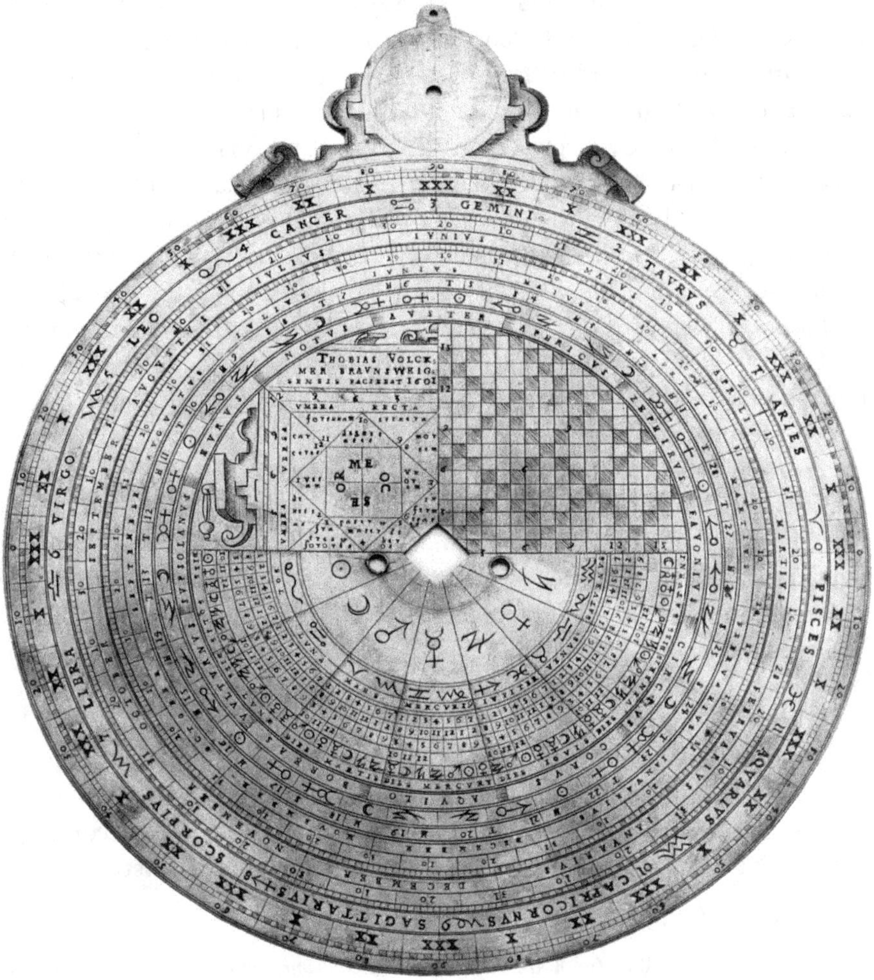

FIGURE 5.4 Astrolabe, by Tobias Volckmer the Elder, 1601; back (Stift Seitenstetten, Austria). © CHRISTINE CISAR, HAAG, AUSTRIA.

signs themselves, zodiacal dignities, elemental qualities, among many others. Usually such instruments provide important data and interpretative tools for horoscopic work, and for these devices the term 'tabular instrument' may be coined. Precursors of such tabulations can be found in astronomical tables of the late Middle Ages,[8] but already by the tenth century a circular display of the

8 Fritz Saaby Pedersen, *The Toledan Tables: A Review of the Manuscripts and the Textual Versions with an Edition*, "Det Kongelige videnskabernes selskab: Historisk-filosofiske Skrifter 24", Copenhagen, C.A. Reitzels Forlag, 2002, 4 vols.; v. 4, pp. 1590–1595.

planetary dignities could be found in the *zij* of al-Battānī.[9] A similar arrangement can be found on an astrolabe by 'Abd al-Karīm al-Miṣrī of 1227–1228.[10]

A rather unusual and hitherto unknown specimen preserved in Berlin consists of two hexagonal, hinged gilt-brass tablets (*Fig. 5.5*).[11] The engraved inscription, "HINRICVS RANTZOVIVS FIERI CURAVIT ANNO D[OMI]NI 1574 ANNO AETATIS, 49", and the escutcheons of the Rantzau family and of Christina von Halle (Heinrich Rantzau's wife) indicate that the tablets were made in 1574. To historians of science Count Heinrich Rantzau (1526–1598), Danish governor of Schleswig-Holstein, is mostly known as the noble host of Tycho Brahe, who stayed in Wandsbeck castle after leaving Denmark in 1597. His father, Johann Rantzau (1492–1565), was an influential politician and general who had witnessed Luther's appearance at the Imperial Diet of Worms in 1521. Deeply impressed, he joined the Protestant camp and sent his son to study in Wittenberg. During a long stay of nearly ten years young Heinrich was influenced by Philipp Melanchthon (1497–1560), an ardent believer in astrology who had studied in Tübingen under the eminent astronomer Johannes Stöffler (1452–1531). Heinrich Rantzau was just one of Melanchthon's many students who left Wittenberg with a staunch belief in the new religious faith of Protestantism and the old art of astrology.[12] With his passion for books, his patronage in various fields of humanistic learning and art, he was considered a 'black swan' among contemporary noblemen.

Broadly speaking, the contents of the first book of Claudius Ptolemy's *Tetrabiblos* are condensed on the outer faces of the tablets. On the right hand side is a table displaying the essential planetary dignities according to Ptolemy.[13] There were five essential planetary dignities in Hellenistic astrology,

---

9 Carlo Alfonso Nallino (ed.), *Al-Battānī sive Albatenii Opus Astronomicum*, "Pubblicazioni del Reale Osservatorio di Brera in Milano 40", Milan, Ulrico Hoepli, 1899–1907, 3 vols.; v. 2, pp. 299 and 309–314.

10 Museum of the History of Science, Oxford, inv. no. 37148.

11 Kunstgewerbemuseum, Berlin, inv. no. K 4682; acquired in 1848 by the Royal Prussian Cabinet of Curiosities; dimensions: 125 mm (width) × 195 mm (height).

12 Günther Oestmann, *Heinrich Rantzau und die Astrologie: Ein Beitrag zur Kulturgeschichte des 16. Jahrhunderts*, "Disquisitiones Historiae Scientiarum: Braunschweiger Beiträge zur Wissenschaftsgeschichte 2", Braunschweig, Braunschweigisches Landesmuseum, 2004, pp. 30–139.

13 Wilhelm Knappich, "Geschichte der zodiakalen Würden", *Zenit: Zentralblatt für astrologische Forschung* 4 (1933), pp. 195–204, 235–244, 273–283 and 336–345; Stephan Heilen, "Ptolemy's Doctrine of the Terms and its Reception", in Alexander Jones (ed.), *Ptolemy in Perspective: Use and Criticism of his Work from Antiquity to the Nineteenth Century*, "Archimedes: New Studies in the History and Philosophy of Science and Technology 23", Dordrecht etc., Springer, 2010, pp. 45–93.

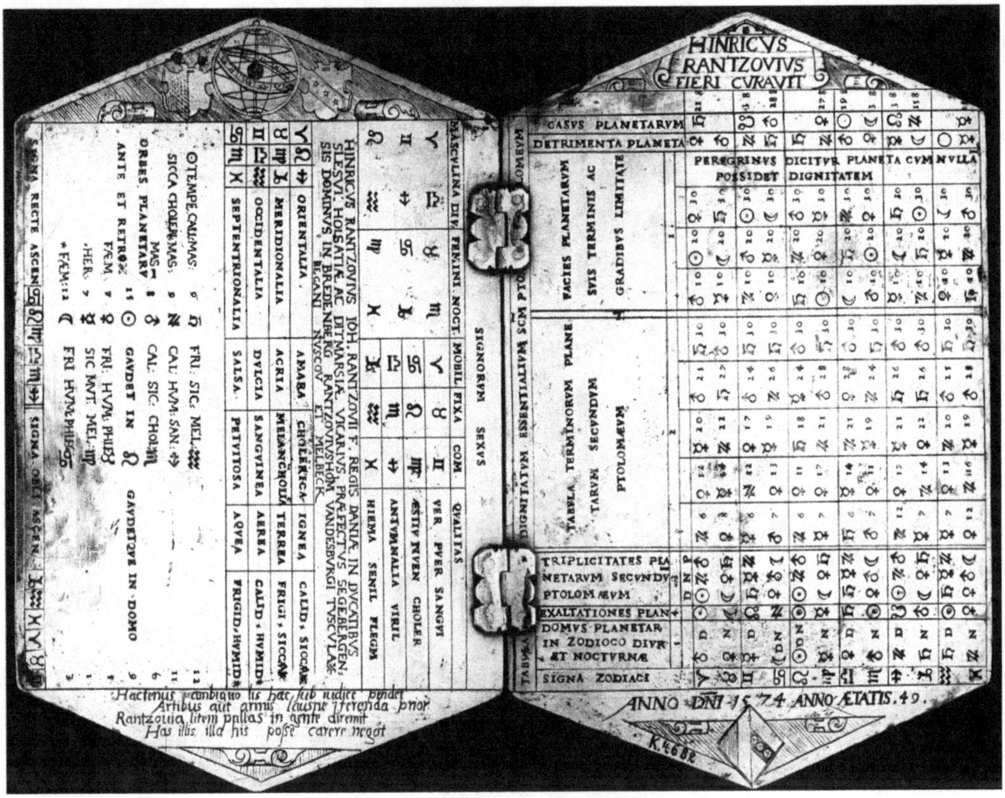

FIGURE 5.5 Astrological tablet owned by Heinrich Rantzau, 1574; outer faces. © STAATLICHE MUSEEN PREUSSISCHER KULTURBESITZ BERLIN, KUNSTGEWERBEMUSEUM.

as described by Ptolemy: houses, triplicities, exaltations, terms (literally: *fines*, boundaries) and faces (*prosopa*).[14] Dignities generally indicate the planet's strength, but have to be distinguished: 'essential' dignities are concerned with the parts of the zodiac to which a fundamental relationship with the planet itself has been ascribed, whereas 'accidental' dignities (or 'debilities') indicate the strength (or weakness) that a planet gains from its accidental location in the zodiac, its position relative to the horizon and the four 'cardines' (Ascendant, Midheaven, Descendant and Lower Heaven), the Sun and its interaction with

14 Claudius Ptolemy, *Tetrabiblos*, ed. and trans. by Frank Egleston Robbins, "Loeb Classical Library 435", Cambridge MA, Harvard University Press, 1940 (reprint: London, William Heinemann Ltd., 1980), pp. 79–113.

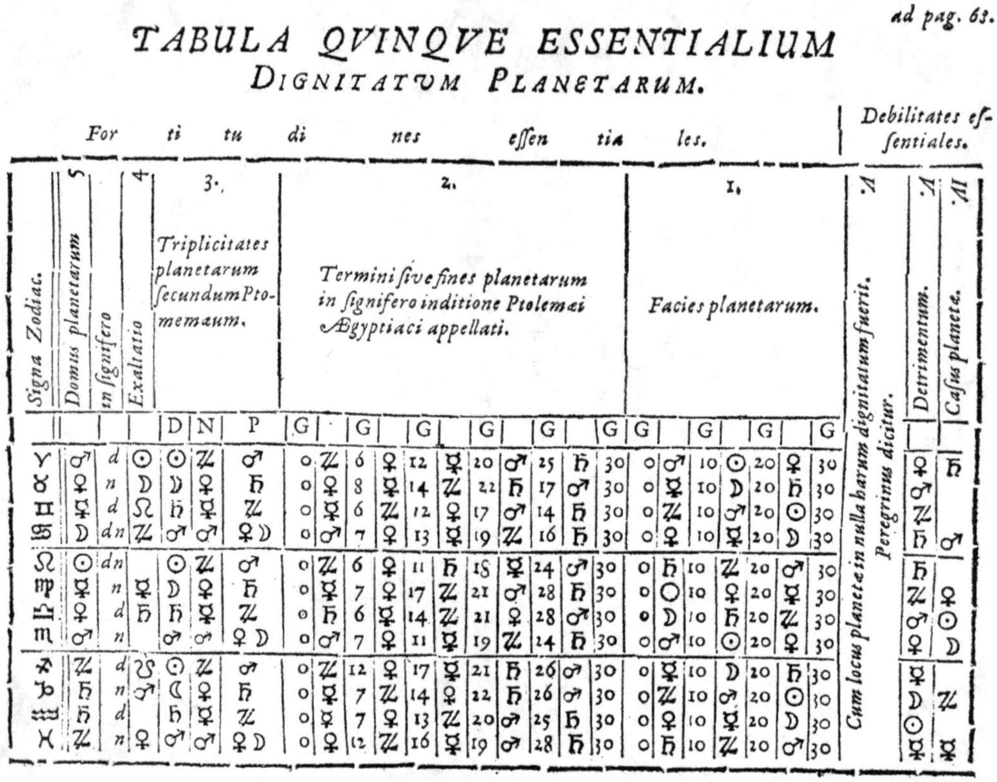

ad pag. 63.

TABULA QVINQVE ESSENTIALIUM DIGNITATVM PLANETARUM.

For ti tu di nes essen tia les. | Debilitates essentiales.

Signa Zodiac.	Domus planetarum (5)	in signifero	Exaltatio (4)	Triplicitates planetarum secundum Ptolemaeum (3.) D	N	P	Termini sive fines planetarum in signifero inditione Ptolemaei Aegyptiaci appellati (2.) G		G		G		G		G		G	Facies planetarum (1.) G		G		G		G	Cum locus planetae in nulla harum dignitatum fuerit. Peregrinus dicitur. (V.)	Detrimentum (V.)	Casus planetae (VI.)
♈	♂	d	☉	☉	♃	♂	0	♃	6	♀	12	☿	20	♂	25	♄	30	0	♂	10	☉	20	♀	30		♀	♄
♉	♀	n	☽	☽	♀	♄	0	♀	8	☿	14	♃	22	♄	17	♂	30	0	☿	10	☽	20	♄	30		♂	
♊	☿	d	☊	♄	☿	♃	0	☿	6	♃	12	♀	17	♂	14	♄	30	0	♃	10	♂	20	☉	30		♃	
♋	☽	dn	♃	♂	♂	♀ ☽	0	♂	7	♀	13	☿	19	♃	16	♄	30	0	♀	10	☿	20	☽	30		♄	♂
♌	☉	dn		☉	♃	♂	0	♃	6	♀	11	♄	18	☿	24	♂	30	0	♄	10	♃	20	♂	30		♄	
♍	☿	n	☿	☽	♀	♄	0	☿	7	♀	17	♃	21	♂	28	♄	30	0	○	10	♀	20	☿	30		♃	♀
♎	♀	d	♄	♄	☿	♃	0	♄	6	☿	14	♃	21	♀	28	♂	30	0	☽	10	♄	20	♃	30		♂	☉
♏	♂	n		♂	♂	♀ ☽	0	♂	7	♀	11	☿	19	♃	24	♄	30	0	♂	10	☉	20	♀	30		♀	☽
♐	♃	d	☋	☉	♃	♂	0	♃	12	♀	17	☿	21	♄	26	♂	30	0	☿	10	☽	20	♄	30		☿	
♑	♄	n	♂	☾	♀	♄	0	☿	7	♃	14	♀	22	♄	26	♂	30	0	♃	10	♂	20	☉	30		☽	♃
♒	♄	d		♄	☿	♃	0	☿	7	♀	13	♃	20	♂	25	♄	30	0	♀	10	☿	20	☽	30		☉	
♓	♃	n	♀	♂	♂	♀ ☽	0	♀	12	♃	16	☿	19	♂	28	♄	30	0	♄	10	♃	20	♂	30		☿	☿

FIGURE 5.6 The essential dignities of the planets, from Cyprianus Leovitius, *Brevis et perspicua ratio iudicandi genituras, ex physicis causis et vera experientia extructa*, London, 1558.

other planets.[15] Although accidental dignity is considered more significant than essential dignity in the interpretation of a chart, both are important.

The same tabular arrangement can be found in the first edition of Rantzau's *Tractatus astrologicus*.[16] He most likely derived it either from Johannes Schöner's *Opusculum astrologicum* or the *Isagoge Astrologiae iudiciariae* by Cyprian Leowitz (*Fig. 5.6*).[17]

15 W. Knappich, "Die akzidentellen Würden der Planeten", *Zenit: Zentralblatt für astrologische Forschung* 5 (1934), pp. 367–376, 397–402.

16 Heinrich Rantzau, *Tractatus astrologicus de genethliacorum thematum iudiciis pro singulis nati accidentibus*, Frankfurt, Joannem Wechelum, 1593, table between pp. 24–25.

17 Johannes Schöner, *Opusculum Astrologicum*, trans. and ed. by Robert Hand, "Project Hindsight: Latin Track 4", Berkeley Springs, Golden Hind Press, 1996, p. 33; Johannes Schöner, *Opera mathematica*, Nuremberg, 1551, fol. 32r; Cyprian Leowitz, *Brevis et*

The table starts with the twelve zodiacal signs on the left (*Signa Zodiaci*), followed by the planetary houses (*Domus planetarum*; these should not be confused with the twelve houses connected with different aspects of a person's life). Each planet (with the exception of Sun and Moon) has a day and a night house, marked *D* (*dies*) und *N* (*nox*). Here a planet assumes rulership and thus becomes host of the sign. It is the most significant essential dignity.

Exaltation (*Hypsoma*) denotes a single degree of the zodiac, in which the respective planet exerts its full power. Diametrically opposite lies a point of depression (*Casus, Deiectio, Tapeinoma*), which is denoted in the rightmost column (its value in ecliptic longitude lies opposite the planet's exaltation and has to be transferred to column 1; e.g. *casus* (depression) of the Sun is 19° Libra, with its exaltation at 19° Aries). Here also the lunar nodes (*Caput Draconis* and *Cauda Draconis*) are noted, i.e. they are classified as planets, as they are in some non-European conceptions of astrology.[18]

According to the doctrine of triplicities, four equilateral triangles can be inscribed within the zodiac, each corresponding to one of the four elements, the sexes and winds. In column four (from the left) the triplicity rulers of day- and night-time events are listed. A third planet (denoted *P*) is also designated as companion or joint ruler. The sequence of these joint rulers as set out by Schöner/Rantzau differs, however, from that used by Ptolemy.

The disposition of terms (*Tabula terminorum planetarum secundum Ptolomaeum*; limits, *horia*) is a fine division of the zodiac between the five planets (Sun and Moon excluded) into sectors of influence of differing size. By Ptolemy's time at least three different and inconsistent versions had been employed,[19] namely "Egyptian", "Chaldean" and a system of his own. The distribution on the tablet follows the Egyptian system, with the respective termini of influence indicated.

The so-called faces of the planets (*facies, prosopon*) following in the next column are of least importance with regard to the astrological interpretation of a chart.[20] This is another fine division of each zodiacal sign into three equal segments of 10°. These 36 decans of the ecliptic were originally assigned to the Egyptian decan gods. Here the planets show their "face".

---

perspicua ratio iudicandi genituras, ex physicis causis et vera experientia extructa, London, Henry Sutton, 1558, fol. EIIIv.

18 Willy Hartner, "The Pseudoplanetary Nodes of the Moon's Orbit in Hindu and Islamic Iconographies", in Hartner, *Oriens-Occidens*, "Collectanea 3", Hildesheim, Olms, 1968, pp. 349–404.

19 Ptolemy, op. cit. (n. 14), pp. 91–107.

20 Robert Albrecht Böker, "Prosopon", in *Paulys Realencyclopädie*, op. cit. (n. 1), v. 23, cols. 869–878; Auguste Bouché-Leclercq, *L'Astrologie grecque*, Paris, Leroux, 1899, pp. 215ff.

The small numbers below the headings denote the significance of the planetary dignities to ascertain the ruler of the chart: house = 5; exaltation = 4; triplicity = 3; term = 2; face = 1. This information is important when evaluating the almuten (the most dignified planet in the chart) by assigning numerical scores to each planet.

On the left side, the astrological qualities of the zodiacal signs are noted. Each sign is assigned to one of the four humours, seasons, elements and temperaments. Each is also designated masculine or feminine, fixed or mutable, and daily or nightly. The names of the months are accompanied by mnemonic verses concerning the labours associated with each.

In the lowest field, the ranges of influence of the seven planets (*orbes planetarum*) are indicated. This is the distance on either side of an aspect, within which the influence of a planet is still partially effective (introducing a certain range is necessary because planets rarely form precise aspects). On the right the accidental dignity of "Joy" is noted, i.e. places in the zodiac where planets "rejoice" and thus have their most beneficial effect.

Finally, the zodiacal signs with straight and oblique ascension are listed. This is very general information, however, entirely insufficient for casting a horoscope. To get more specific information one has to unfold the tablets (*Figs. 5.7* and *5.8*). The inner faces show the rising times of the zodiacal signs, the true place of the Sun, and the length of day and night calculated for the latitude of 55° (the latitude of Breitenburg, Rantzau's principal castle northwest of Hamburg, which was often placed about 1° too far north in the sixteenth century).[21] Each table covers half the year, with the vernal equinox on the left and the autumnal equinox on the right. On the left margin there are 24 hours (divided into day and night), followed by six columns for rising times, names of the months, date, zodiacal signs, length of day and night, and times for sunrise and sunset.[22] As well as these astronomical parameters, the tables provide the method of house division according to Regiomontanus. This so-called "rational method" (in which circles of position joining at the north and south point of the observer's horizon are laid at distances of 30° around the celestial equator, thus giving unequal divisions of the ecliptic) was widely used by astrologers in the early modern period. Instructions for the use of the tables have been preserved in a manuscript by Rantzau (see *Appendix*).[23]

---

21 For further details see Oestmann, op. cit. (n. 12), pp. 123–124.

22 Identically arranged tables for a latitude of 55° can be found in Rantzau, *Diarium sive Calendarium Romanum, oeconomicum, ecclesiaticum, astronomicum, et ferè perpetuum*, Wittenberg, Christophorus Axinus, 1593, pp. 168–190.

23 Vienna, Austrian National Library, Cod. Vin. 11449, fols. 247r–248r. For the original text and German translation, see Oestmann, op. cit. (n. 12), pp. 84–85.

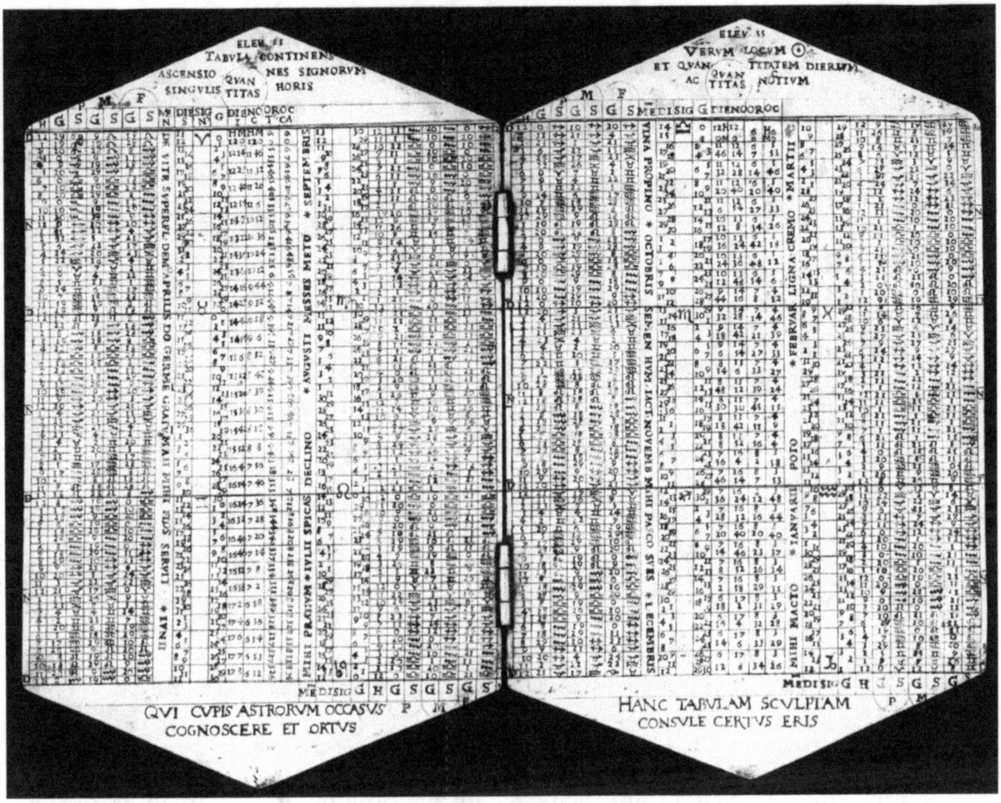

FIGURE 5.7 Astrological tablet owned by Heinrich Rantzau, 1574; inner faces. © STAATLICHE MUSEEN PREUSSISCHER KULTURBESITZ BERLIN, KUNSTGEWERBEMUSEUM.

The design of the tablets is a testimony to Rantzau's aspirations for a thoroughly organized and schematized arrangement of astrological doctrine, which he later displayed in his *Tractatus astrologicus*. Such tabular arrangements were also executed on other instruments. In 1557, the French physician Claude Dariot (1533–1594) described an astrological disc on which the essential dignities of the planets were marked (see *Fig. 4.5* in Stephen Johnston's chapter). The back of this instrument showed "the degrees Masculine and Feminine, bright, smokie, and darke, vacant or voyde, deepe, augmenting Fortune, and diminishing Fortune, and of those which bring imbecilitie or weakenesse to the body which are called Azemenes".[24] An astrological disc by Gerard

24 Claude Dariot, *A breefe and most easie Introduction to the Astrologicall iudgement of the Starres: Whereby euery man maye with small labour giue aunswere to any question*

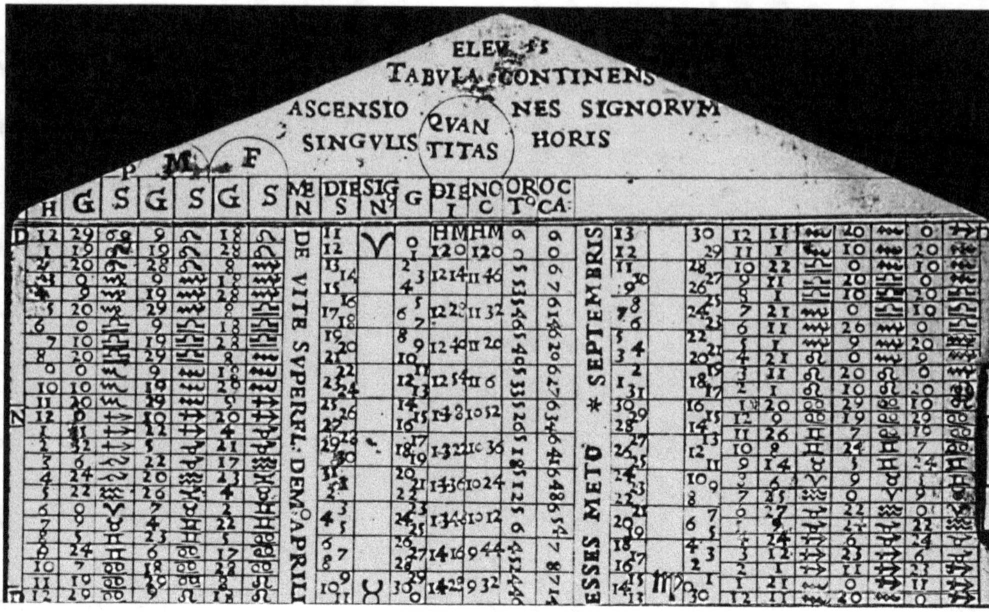

FIGURE 5.8 Astrological tablet owned by Heinrich Rantzau, 1574; inner face (detail). © STAATLICHE MUSEEN PREUSSISCHER KULTURBESITZ BERLIN, KUNSTGEWERBEMUSEUM.

Mercator discovered some years ago displays the nature and qualities of the zodiacal signs, the essential planetary dignities, divisions into male and female degrees, as well as luminous, dark/shadowy, smoky or void degrees of a sign.[25] Similar scales are also found on the back of an astrolabe probably made for King Edward VI, which has been attributed to Thomas Gemini (*c.* 1515–1562).[26] A connection between these two instruments seems likely, with Mercator's

demaunded, London, n.d. [*c.* 1583], unpaginated. The original Latin edition was published in 1557 in Lyon under the title *Ad astrorum iudicia facilis introductio*, with an edition in French appearing the following year in Lyon under the title *L'Introduction au iugement des astres avec un Traité des elections propres pour le commencement des choses*.

25 Historical Museum, Basle, inv. no. 1876–20; see Steven Vanden Broecke, "Dee, Mercator, and Louvain Instrument Making: An Undescribed Astrological Disc by Gerard Mercator (1551)", *Annals of Science* 58 (2001), pp. 219–240.

26 National Maritime Museum, Greenwich, AST0567; see George H. Gabb, "The Astrological Astrolabe of Queen Elizabeth", *Archaeologia or Miscellaneous Tracts Relating to Antiquity* 86 (Ser. II, 36) (1936), pp. 101–103; Gerard L'E. Turner, *Elizabethan Instrument Makers: The Origins of the London Trade in Precision Instrument Making*, Oxford, Oxford University Press, 2000, pp. 95–97.

disc probably serving as a template for the astrolabe.[27] This instrument had an explicitly didactic function, which is described on the back by Mercator:

> To be sure, the instrument is not able to embrace all things, and we wanted no more than to propose the principal headings of the whole art in such a way that they could lead to her innermost parts and would be a most powerful summary and reminder of things scattered among authors. When he has understood these things, [the student of astrology] should set, for a given time, the indices of the individual planets in that place where they were found from the ephemerides or by some other means; it will immediately be apparent how powerful each [planet] is, and which is the most powerful on account of the part of the ecliptic in which it stands. [...] Thus he will have set up very suitably a figure of the sky for that time and place and will openly contemplate all things in one gaze. There will be no more need for lengthy wandering among the authors or a good memory of many things. With the nature set down from physical explanations under every single index, by comparing them with one another as well as with others set forth on the outside, he will easily indicate what any planet signifies, what strength it has, what advantage or disadvantage it receives elsewhere and in what order the planets mutually succeed each other in power. But he who is not a natural philosopher has to follow the writings of the astrologers and, as if in an immense forest, gets lost amidst a multitude of meanings.[28]

27 The link may have been John Dee, who stayed in Leuven in 1547–1550 and was in close contact with Mercator; see Gerard L'E. Turner and Koenraad Van Cleempoel, "A Tudor Astrolabe by Thomas Gemini and its Relationship to an Astrological Disc by Gerard Mercator of 1551", *The Antiquaries Journal* 81 (2001), pp. 400–409. For Gemini's biography, see Turner, op. cit. (n. 26), pp. 12–20.

28 "[...] neque enim omnia potuit instrumentum capere, neque nos aliud voluimus quam praecipua totius artis capita sic proponere, ut ducere possint ad eius intima, & rerum apud scriptores sparsarum collectio & monumentum essent praesentissimum. Haec cum tenurit, statuat pro dato aliquo tempore singulorum planetarum indices eo loco quo ephemeridibus aut alia quavis via reperti fuerint planetae mox apparebit quam quisque sit potens, & quis potentissimus existat ratione partis eclypticae in qua positus est. [...] sic iam figuram coeli pro eo tempore & loco aptissime erexerit, coramque omnia simul uno intuitu contemplabitur, nec longa per autores vagatione, neque alta multarum rerum memoria opus erit, sed ex Physicis rationibus naturas sub unoquoque indice positas, conferendo tam inter se mutuo quam cum alijs extrapositis, facile indicabit quid quisque planeta significet, quid valeat, quid aliunde commodi aut incommodi accipiat, quo ordine sibi mutuo in potentia succedant Planetae, qui vero Physicus non fuerit, necesse est Astrologorum scripta sequatur, & per multitudinem significationum velut per immensam sylvam oberret"; Instruction sheet on the back of Mercator's astrological disc; English translation from Vanden Broecke, op. cit. (n. 25), pp. 234 and 239.

Conclusion

The huge variety of meanings and attributions bound up in astrological interpretation could be conveniently handled by using tabular instruments. Even a well-versed astrologer was not able to recall all these figures and details from memory, so these small and easily manageable tablets must have been very useful as an aide-mémoire for Rantzau when casting and interpreting horoscopes.

Appendix: Description of Heinrich Rantzau's Astrological Tablet

Vienna, Austrian National Library, Cod. Vin. 11149, fols. 247r–248r.

[Fol. 247r] Declaratio Tabulae Mathematicae quae complectitur ascensiones signorum Zodiaci ad singulas diei et noctis horas per totum annum. Haec tabula divisa est in duplicem faciem, sinistram et dextram, quarum quaelibet in tres ordines est distributa. Primus ordo sinistrae faciei in fronte tres lit[t]eras maiusculas P. M. et F. scriptas habet, quarum P principium, M medium et F finem signi significat. Distinguitur primus ordo in septem columnas. Harum prima horas diei continet, eademque in sinistro latere D et N lit[t]eras diei sive noctis horas significantes alternatim ascriptas habet. Sequentes vero sex columnae, sub lit[t]eris P. M. et F. collocatae ascensiones signorum Zodiaci in gradibus ad singulas horas complectuntur, atque ex his sex, primae duae columnae principio signi, sequentes duae medio, ac ultimae duae fini signi inserviunt. Ut exempli gratia, cupio scire, quodnam signum Zodiaci et quotus eius gradus 1 [sic; 11] Martii ad horam eius diei 12 ascendat *supra* horizontem: Ingredior itaque hanc tabulam et hora praedicta in summitate primae columnae inventa sub lit[t]era Maiuscula, P quae principium signi significat, deprehendo ascendere 29 gradum Cancri.

II. Ordo intermedius huius sinistrae faciei itidem undecim columnas dividitur, quarum prima [fol. 247v] menses, secunda vero numerum dierum mensis, tertia Zodiaci signa Aries, Taurus, Gemini, quarta gradus signorum, quinta et sexta quantitatem dierum ac noctium proponunt atque hae duae columnae duplicem numerorum ordinem habent. Alter gradus [sic; horis] continet, alter minuta, quod lit[t]eris G. [sic; H] et M. in fronte scriptis significatur. Septima et octava ortum et occasum Solis in gradibus [sic; horis] et minutis, Nona aliquot menses, decima mensium dies, undecima tria Zodiaci signa Cancer, Leo, Virgo, duodecima gradus signorum oculis exponit. Exempli gratia: cupio scire introitum Solis in signum aliquod Zodiaci, item verum locum Solis ad 12 diem

Mart[ii] praeterea quantitatem diei et noctis, nec non ortum et occasum solis. Ingressus itaque hanc tabulam, deprehendo ad praedictum diem, quem ingredi gradum 1 Arietis, quantitatem eius diei esse in horarium noctis totidem, oriri solem hora sexta et occidere eadem. Atque hic ordo tanquam intermedius reliquis duobus utriusque inservit. Tertius ordo huius sinistrae faciei in 7 columnas discernitur et non secus atque primus ascensiones signorum Zodiaci singulis horis exhibet. Nam ubi primus ordo desinit, ibi hic incipit, nihilque ab eo differt, nisi quod huius ratio ab inferiore parte incipiens sursum progrediatur. Quod eo consilio provisum est, ne haec tabula iustam excedat longitudinem. Atque haec de sinistra tabulae facie dixisse sufficiat.

[fol. 248r] Dextra facies a sinistra nihil prorsus differt, nisi quod illa fere tantum primis sex signis haec vero reliquis deserviat.

Ortus Lunae ex hac tabula hoc modo etiam potest inveniri: Vide in Ephemeridibus, quod signum Luna ad meridiem diei obtineat et simul considera gradum oppositum, quo facto quaere illud signum in prima columna signorum et iuxta illud signum, horas appositas et facta proportione a meridie, habes ortum Lunae.

Quin et figurae nativitatis ex hac tabula exigi possunt secundum modum Rationalem: Nam hae iuxta modum rationalem exstruuntur secundum aequatorem et aequator habet gradus 360. Quadrans autem caeli habet 90 gradus et singulis horis ascendunt de aequatore 15 gradus. Invento itaque ascendente ab eo numera sursum duas horas et habes duodecimam domum, deinde quatuor et habes 11, postea sex et habes 10 domum. Deorsum numera 2 horas et deprehendes secundam domum, porro numera 4 et habes 3 domum. Si autem horis temporis tui oblati adhaerent minuta facta proportione inter duas horas immediate sequentes provenit figura iuxta modum rationalem. His inventis signa opposita per se patent.

Translation

Explanation of the mathematical tablet, which includes the ascensions of the signs of the zodiac for the individual hours of the day and night for the whole year. This tablet is divided into two parts, a left and a right side, each of which is divided into three rows. The first row on the front left has three capital letters, P, M, and F, of which P marks the beginning [0°], M the middle [15°] and F the end [30°] of a sign. The first row is divided into seven columns. Of these the first contains the hours of the day and bears the letters D and N on the left side, indicating the hours of the day or night, added alternately. The following six columns are placed below the letters P, M and F, and record the ascensions of the signs of the zodiac in degrees at single hours. Of those six, the first two columns are intended for the beginning of the sign, the following two for the middle and the last two for the end of a sign. An example: I want to know, which zodiacal sign and which degree of it ascends above the horizon on the 1st [sic, 11th] of March at 12 o'clock. I enter the table, and when I have found the above-mentioned time at the top of the first column under the capital letter P, which signifies the beginning of the sign, I find that the 29th degree of Cancer is rising.[29]

II. The middle row of the left side is also divided into eleven columns, of which the first indicates the months, the second the number of days of the months, the third the signs of Aries, Taurus and Gemini, the fourth the degrees of the signs, the fifth and sixth the length of days and nights. These two columns have a double sequence of numbers. One contains the hours, the second the minutes, which are indicated by the prefixed letters G [this should read correctly H] and M. The seventh and eighth column indicate the rising and setting of the Sun in hours and minutes, the ninth several months, the tenth the day of the month, the eleventh the three signs Cancer, Leo, Virgo, and the twelfth displays the degree of the signs. An example: I want to know in which sign of the zodiac the Sun enters, as well as the true position of the Sun on March 12th, also the length of day and night and equally the time of the rising and setting of the Sun. Thereupon I enter the table and find for the aforementioned day that the Sun enters the first degree of Aries, that day and night are of equal length, and that the Sun rises and sets at 6 o'clock. And this middle row serves the other two on each side. The third row on the left side is divided into seven columns and—like the first row—displays the ascensions of the signs at individual hours. Where the first row ends, this one begins and does not differ at all from it, except that its arrangement proceeds from the bottom upwards. This is provided intentionally, so that this tablet should not exceed an appropriate length. And it may suffice having said this about the left side of the tablet.

The right side differs from the left only insofar, as it generally serves the first six signs only, and the latter serves the remaining [signs].

The rising of the Moon can also be found with this tablet as follows: Find in the ephemeris which sign the Moon occupies at noon of the day and at the

29 Column P [*Principium*] denotes 0–9°, M [*Medium*] 10–19°, and F [*Finem*] 20–29° of a sign. The position of the Sun in the zodiac for the given date is taken from an ephemeris and the appropriate column for the determination of the ascendant is chosen. On 11 March (Julian calendar), the Sun is in 0° Aries, and the ascendant at 12:00 in column P is 29° Cancer. On 23 March the Sun is in 12° Aries, and correspondingly the ascendant at noon can be found in column M (9° Leo).

same moment view the degree opposite. Then look for that sign in the first column of signs and for the hours marked besides that sign, and if you have calculated the proportion [the fraction of the time] from noon, you will get the Moon's ascension.

Even nativities can be calculated with this tablet according to the rational method [of Regiomontanus]: These are erected in a rational manner corresponding to the [celestial] equator, and the equator has 360 degrees. The quadrant of the sky has 90 degrees, and in a single hour 15 degrees of the equator rise. Therefore, if you have found the ascendant, count two hours from it upwards, and you have the twelfth house, then four, and you have the eleventh, then six and you have the tenth. From there count two hours downwards, and you will find the second house, then four and you have the third house. However, if the hour [of your birth] includes minutes, you can obtain the [horoscope] figure according to the rational method by proportion [i.e. interpolation] between two immediately subsequent hours. If this has been found, the opposite signs are immediately clear.

CHAPTER 6

Astrological Aspectaria on Early Modern Instruments

Louise E. Devoy

During the early modern period European makers produced a huge range of astronomical timekeeping instruments.* Surviving examples have shown us how master craftsmen sought to demonstrate their skill by producing increasingly complex devices filled with perpetual calendars, nocturnals, epact tables, tide calculators, magnetic compasses, wind vanes and latitude tables.[1] Hidden among these gentlemanly tools of knowledge and calculation one can often find an astrological diagram, generally overlooked in the historical literature, namely the aspectarium. Usually found within the inner circle of the lunar volvelle, the aspectarium depicts the geometric relationship between the Sun, Moon and planets within a horoscope, with the angles signifying both positive and negative influences upon a person's fate.

This chapter seeks to unpick the elusive presence and significance of these diagrams through a review of over sixty instruments held in major museum collections in the United Kingdom and a few select items elsewhere. Investigating the results more closely reveals several trends about the different types of aspectaria used, their possible origin, the makers who chose to include them in their work, and whether the diagrams were functional or purely decorative. This analysis will contribute to our continuing understanding of the relationship between astrological practice and instrument-making during the early modern period.

* I am very grateful to Oliver Cooke (British Museum) and Sophie Waring (formerly of the Museum of the History of Science, Oxford), as well as my colleagues at Royal Museums Greenwich, who facilitated access to the objects and images required for this survey. Comments and suggestions made by participants at the XXXIV SIC Symposium in Turin, 7–11 September 2015, were also immensely helpful. I would like to express my special thanks to Marisa Addomine, Jane Desborough, Richard Dunn, Richard Kremer and Josefina Rodríguez-Arribas.

1 For a general overview of these instruments see Anthony Turner, *Early Scientific Instruments: Europe 1400–1800*, London, Sotheby's Publications by Philip Wilson Publishers, 1987, and Gerard L'Estrange Turner, *Elizabethan Instrument Makers: The Origins of the London Trade in Precision Instrument Making*, Oxford, Oxford University Press, 2000.

 | DOI:10.1163/9789004381438_008

The History of Astrological Aspects

Astrological aspects are defined by the relative positions of the Sun, Moon and planets in relation to the ecliptic, or circle of the zodiac signs.[2] When two of these celestial bodies (sometimes referred to as "lights" or "luminaries") are positioned at specific angles, say at 60° or 90° apart, they are said to be in aspect. For astrologers, these geometrical relations can have a positive or negative influence within an astrological figure. The term "aspect" is derived from the Latin *aspicere*, meaning "to regard"; other meanings encompass a sense of the planets "viewing" or "witnessing" each other.[3]

The first written evidence concerning planetary aspects can be found in Ptolemy's *Tetrabiblos* (second century AD), which brought together Babylonian omen literature and astronomical data with the Egyptian divisions of the sky to create a more mathematical form of astrology.[4] Astronomers and astrologers divided the ecliptic circle into 360°, subdividing again into twelve sections of 30° for each zodiac sign. For the simplest aspects, Ptolemy describes how at opposition (180°) the planets face each other along a straight line. Taking inspiration from intervals in music, he then explains how the full circle of 360° can be divided by two, three, four and six to create the aspects of opposition, trine, quartile and sextile. Although Ptolemy does not specifically mention conjunction as an aspect, this direct alignment between planets (0°) was later adopted as the fifth aspect.

As a planet approaches an aspect it is said to be *applying*; as it moves away it is said to be *separating*.[5] Ptolemy describes how each aspect is assigned a positive or negative influence:

> Of these aspects trine and sextile are called harmonious because they are composed of signs of the same kind, either entirely of feminine or

2 For an overview of how aspects are used by contemporary astrologers, see Nicholas Campion, *The Ultimate Astrologer: A Simple Guide to Calculating and Interpreting Birth Charts for Effective Application in Daily Life*, London, Rider, 2002, pp. 93–115.

3 Deborah Houlding, "The Classical Origin & Traditional Use of Aspects", www.skyscript.co.uk/aspects.html (accessed 13 April 2018).

4 William J. Tucker, *Ptolemaic Astrology: A Complete Commentary on the* Tetrabiblos *of Claudius Ptolemy*, Delhi, Astrological Publications, 1962, p. 51; Tamsyn Barton, *Ancient Astrology*, London, Routledge, 1994, p. 30.

5 Claudius Ptolemy, *Tetrabiblos*, ed. and trans. by Frank Egleston Robbins, "Loeb Classical Library 435", Cambridge MA, Harvard University Press, 1940 (reprint: London, William Heinemann Ltd., 1980), (I, 24), pp. 112–113.

> entirely of masculine signs; while quartile and opposition are disharmonious because they are composed of signs of opposite kinds.[6]

He then explains how the planets have certain powers in specific signs of the zodiac which affect their influence within an aspect. For example, Venus is said to be in "exaltation" or greatest effect in Pisces but has her "depression" or least effect in Virgo. Hence one can generate multiple aspects from the different variables involved: aspect shape, zodiac sign, direction of travel of each planet (applying or separating) and the combination of luminaries, i.e. Sun and Venus, Sun and Mars, Mars and Saturn, and so forth.[7] In the centuries after Ptolemy, planetary aspects continued to be included in astrological handbooks, such as the *Matheseos Libri VIII* (Eight Books on the Theory of Astrology), produced by the retired Sicilian lawyer, Julius Firmicus Maternus (fourth century AD). In the second book, Firmicus introduces the aspects as "among the most important things which we must discuss", and describes the Ptolemaic aspects and their influences, including the "unaspected", i.e. conjunction.[8] Yet shortly after Firmicus embarked on his writings in AD 334, astrology came under attack from the authorities under Constantine I (*c.* 272–337, r. 306–337) and subsequent emperors, who regarded it as incompatible with the Christian faith and imposed harsher penalties on practitioners. As a result, the number of surviving astrological texts from the ensuing centuries dwindled. Thankfully, some elements of Firmicus' work survived and references to his book began to emerge in monastic libraries during the eleventh and twelfth centuries, with the first printed version of the *Matheseos* finally appearing in Venice in 1497, reinforcing awareness of the Ptolemaic aspects.[9]

Meanwhile, European scholars were also supplementing their astrological knowledge with the Latin translation of the Arabic text of Al-Qabisi's *Book of the Introduction to the Craft of Astrology*, which was adopted as the main textbook for teaching astrology within medicine during the medieval period. It was originally written around 950 in Aleppo, where Al-Qabisi (later known in Europe as Alcabitius) was a mathematician and astrologer employed at the court of Sayf al-Dawla (916–967, r. 945–967). Alcabitius composed several textbooks on the teaching and testing of court astrologers to ensure the rigour of the profession. His *Introduction* was later translated into Latin and underwent

6 *Ibid.*, (I, 13), pp. 74–75.

7 *Ibid.*, (I, 17–19), pp. 79–91.

8 Firmicus Maternus, *Ancient Astrology, Theory and Practice* (*Matheseos Libri VIII*), trans. by Jean Rhys Bram, New Jersey, Noyes Press, 1975, pp. 53–56.

9 *Ibid.*, pp. 6–7.

twelve printings during the period 1473–1521, with further translations produced in several European languages.[10] Consequently, it was Alcabitius' definition that became the standard:

> The aspects. It is said that one sign looks at another among the signs. This means that each sign looks at the third sign from it in front of it or the third behind it, which is the eleventh from it. This is called sextile aspect and is of friendship. It looks at the fourth sign from it in front of it or the fourth behind it, which is the tenth from it; this is called quartile aspect and is of antagonism, and it is half-enmity. It looks at the fifth from it in front of it or the fifth behind it, which is the ninth from it; [this is called] trine aspect and is of compatibility. It looks at the seventh from it; [it is called] an aspect of opposition and enmity. When planets are in the signs which aspect, they are also called "aspecting" with the same aspect; and when two planets are in one sign, they are called "in conjunction". When a planet is in a sign, its rays are in the sign it aspects, in the same degree and minute of that sign.[11]

Some Latin translations of Alcabitius' work also contained a geometric diagram which clearly demonstrated the aspects in relation to the signs of the zodiac, along with symbols for sextile (✶), quartile (□), trine (△) and opposition (☍).

This corpus of five Ptolemaic aspects remained unchanged until the beginning of the seventeenth century, when Johannes Kepler (1571–1630) created three new aspects based on musical ratios. Kepler introduced these new aspects in *De Fundamentis Astrologiae Certioribus* (Concerning the More Certain Fundamentals of Astrology, 1602), written hastily in Prague after his succession as Imperial Mathematician at the court of Rudolph II, following the death of his predecessor, Tycho Brahe (1546–1601) in October 1601.[12] Kepler described his new aspects as follows:

> Thesis 38: Therefore since there are eight formative ratios of motions, there is some action (or, as it were, motion) of heaven on the earth when it meets an intermediating ray of a star and they then come together on

10 Al-Qabisi (Alcabitius), *The Introduction to Astrology*, trans. by Charles Burnett, Keiji Yamamoto, Michio Yano, London and Turin, Warburg Institute, 2004, pp. 1–5.

11 *Ibid.*, p. 27.

12 Johannes Kepler, *The Harmony of the World*, trans. by E.J. Aiton, A.M. Duncan, J.V. Field, Philadelphia, American Philosophical Society, 1997, pp. xx–xxi.

> earth to form angles; thus the eight harmonic ratios will be expressed in the dimensions of these angles. Now the ancients did not acknowledge more than five [of these angles] (commonly called Aspects): Conjunction, Opposition, Quadrature, Trine, and Sextile. But reason soon led me to add three more: quintile, biquintile, and sesquiquadrature, which many experiences have since confirmed.[13]

Seventeen years later, Kepler expanded on his new theory of the aspects in the *Harmonice Mundi* (On the Harmony of the World, 1619), which was partly inspired by his reading of Ptolemy's *Harmonica* and partly by weather observations that did not correspond with the existing theory of aspects.[14] In the *Harmonice Mundi*, Kepler explains how the aspects are generated by geometrical shapes within the circle of the zodiac rather than using Ptolemy's derivation based on musical ratios. From this analysis, he created thirteen aspects, summarized below in order of decreasing potency:

1. conjunction and opposition (0°, 180°);
2. quadrature (90°);
3. trine, sextile, semi-sextile (120°, 60°, 30°);
4. quintile, biquintile, quincunx (72°, 144°, 150°);
5. decile, tridecile, octile, trioctile (36°, 108°, 45°, 135°).[15]

Kepler illustrates his ideas alongside the text, with the aspects forming increasingly complex shapes inside the circle of the zodiac, culminating in a dodecahedron, or twelve-pointed star. At the end of Book IV, however, Kepler admits that he would discard all minor aspects, "in case the number of aspects should become too great", ending the sequence at biquintile.[16]

While Kepler continued to develop his theory of harmony among the motions of the planets, observations of the sky by other astronomers were also changing astrology beyond the traditional five Ptolemaic aspects. The invention and subsequent development of the telescope led over the next 300 years to the discovery of planets previously invisible to the naked eye. In response, astrologers had to create ever more complex combinations of aspects to

---

13 J. Bruce Brackenridge and Mary Ann Rossi, "Johannis Kepler's *Concerning the More Certain Fundamentals of Astrology*, Prague 1601", *Proceedings of the American Philosophical Society* 123 (1979), p. 97. For the original Latin text see Johannes Kepler, *De Fundamentis Astrologiae Certioribus*, in *Gesammelte Werke. Band IV. Kleinere Schriften 1602/1611 Dioptrice*, Max Caspar and Franz Hammer (eds.), Munich, C.H. Beck, 1941, p. 22.

14 Mary Ellen Bowden, "The Scientific Revolution in Astrology: the English Reformers, 1558–1686", unpublished doctoral thesis, Yale University, 1974, pp. 109–129.

15 Kepler, *The Harmony of the World*, op. cit. (n. 12), pp. xxxii and 340.

16 *Ibid.*, p. 347.

accommodate these influences, although the more distant planets were regarded as only having an influence on long-term historical changes and not on individual horoscopes.[17] Astrologers today use computer programmes to create these complex horoscopes with multiple aspects that bear little relation to the simple geometric diagram of the early modern period, yet one can still recognize the original Ptolemaic principles.

Prevalence of Aspectaria on Instruments

In order to assess the prevalence of aspectaria on early modern astronomical instruments, the author reviewed a sample of over sixty held in four major British museum collections: Science Museum, British Museum, Museum of the History of Science, Oxford and the National Maritime Museum (part of Royal Museums Greenwich).[18] The sample also included two unusual examples from the Museo Galileo in Florence and a unique astrological disc in the Historisches Museum, Basel. Although this is not a complete catalogue of all known examples of aspectaria on instruments, several trends have become apparent in the data, which is summarized in *Appendix 1*. In this table, the objects have been grouped by instrument type (diptych dial, astronomical compendium, nocturnal, etc.) and then chronologically. The remaining columns classify the outline shape of each aspectarium, the symbols and text incorporated into the design and whether the aspectarium could be used in a functional way to determine planetary aspects. These trends form the basis of the following discussion.

Overall, it is clear that the appearance of aspectaria mirrors the chronology of the main production period of these instruments, from the earliest designs in the 1480s through to the 1650s, with a few further examples appearing later in the eighteenth century. The majority of instrument makers who included aspectaria in their designs were based in the main production centres of southern Germany, such as Augsburg, Nuremberg and Munich, with additional interest shown by makers further west in Cologne and east in Dresden. We can also see from the chronological spread of the data that aspectaria were not just a fashionable trend during one period, but continued to feature in the output of successive generations, such as Christoph Schissler (*c.* 1531–1608)

17 Nick Campion, *The Practical Astrologer*, Twickenham, Hamlyn Publishing, 1987, pp. 76–77.

18 These items can be viewed in detail on each museum's online database (see *Appendix 1* for details) or on EPACT, www.mhs.ox.ac.uk/epact/.

and his son, Hans Christoph Schissler junior (*c.* 1561–*c.* 1626), or the Munich-based maker Ulrich Schniep (*fl.* 1545–1588) and his son Alexius (*fl.* 1571–1613).[19]

In terms of instrument type, the astronomical compendium is by far the most common device for containing an aspectarium, which usually fills the central area of the lunar volvelle. This disc, numbered from 1 to 29½ days, with a circular aperture that reveals a pictorial representation of the lunar phases, is generally used to convert between lunar time and solar time when using the sundial as a Moon dial. This volvelle is frequently located on the underside of the top leaf (noted as leaf Ib in *Appendix 1*), apart from a handful of examples, where the aspectarium is prominently displayed on the top of the upper leaf (noted as leaf Ia) as part of a nocturnal, calendar scale or wind vane.[20] For the London instrument makers Humfrey Cole (*fl.* 1568–1591), Charles Whitwell (*fl.* 1582–1611), James Kynvyn (*fl.* 1590–1610) and Elias Allen (*c.* 1588–1653),[21] the lunar volvelle containing the aspectarium was part of a much more complex set of circular zodiacal and calendrical scales (*Fig. 6.1*). This suggests that the diagram may have had a practical function, as discussed below.

The second most common instrument to feature aspectaria is the nocturnal, with the aspectarium generally at the centre of the inner volvelle. More specifically, it is the sixteenth-century metal hybrid nocturnal, where the instrument maker sometimes includes vertical dials or Rojas projections on the other side, which includes an aspectarium, rather than the more mundane boxwood versions produced from the later seventeenth century.

The third most common group is diptych dials, although it is the metal square box type which is more likely to have an aspectarium than the ivory type associated with well-known makers such as Georg Hartmann (1489–1564), Paul Reinmann (*c.* 1557–1609) and the successive generations of the Tucher family (*fl.* 1550–1632).[22] On these ivory instruments, the lunar volvelle is either filled with a stylized flower or a Moon and stars motif.

The main exception to this trend is the Troschel family of Nuremberg, with both the father Hans the Elder (*c.* 1549–1612) and son Hans the Younger (*c.* 1599–*c.* 1634) including aspectaria within the lunar volvelles of their ivory

19 For biographical details, see EPACT, "Makers", www.mhs.ox.ac.uk/epact/makers.php (accessed 13 April 2018).

20 See the following items in *Appendix 1*: MHSO 43055, 49061, 80277, 38997, 49305; BM 1855, 1201.225.

21 See EPACT, "Makers", cit. (n. 19) and Turner, op. cit. (n. 1), for more biographical details on Cole, Whitwell and Kynvyn. For Elias Allen see Hester K. Higton, "Allen, Elias (*c.* 1588–1653)", *Oxford Dictionary of National Biography*, Oxford University Press, 2004, online edition, January 2008, www.oxforddnb.com/view/article/37108 (accessed 15 November 2015).

22 EPACT, "Makers", cit. (n. 19).

FIGURE 6.1 Aspectarium on an astronomical compendium, by Humfrey Cole, 1569 (National Maritime Museum, Greenwich, no. AST0172).
© NATIONAL MARITIME MUSEUM, GREENWICH, LONDON.

diptych dials.[23] Intriguingly, both used a different aspectarium design from that of their Nuremberg counterparts, favouring a functional diagram rather than a decorative version.[24] The only other ivory diptych dials to feature aspectaria were individual examples by Christian Heiden (1526–1576) and an unsigned dial in the collections of the British Museum, believed to have been produced in the Netherlands in 1595.[25] No aspectaria were found on ivory Dieppe magnetic azimuth dials, although there is one on a metal version by the Augsburg clockmaker Johannes Buschmann (*c.* 1632–1676).[26]

The remaining objects listed in *Appendix 1* are an eclectic mix, from perpetual calendars (MHSO 53916 and 78935) to a vertical disc dial (MHSO 33775), a mechanical equinoctial dial (MHSO 41087) and an elaborate astronomical table clock attributed to Caspar Rauber (*fl.* 1570s), which has an aspectarium on the lunar volvelle in the centre of the rete (MG 3370).[27]

It is curious to note that there are few examples of aspectaria on astrolabes, despite their specific design for the purposes of astrology, particularly within the Islamic world. The most similar feature to the aspectarium is the table of exaltations and depressions of the planets, which denotes the points on the ecliptic at which the planets reach their strongest and weakest positions of influence, but again, this has only been recorded on a handful of Islamic and European instruments.[28] There are three exceptions to this general trend. The first is a European astrolabe produced in France around 1630 with an aspectarium inscribed on a lunar volvelle positioned underneath an upper volvelle with a Rojas projection (*Fig. 6.2*). The second is an astrolabe-type astrological instrument used to determine lunar aspects for assessing the most auspicious

---

23 See Bruce Chandler, Clare Vincent, "Three Nürnberg Compassmacher: Hans Troschel the Elder, Hans Troschel the Younger and David Beringer", *Metropolitan Museum Journal* 2 (1969), pp. 211–216.

24 See *Appendix 1*: BM 1888, 1201.287 (Troschel the Elder); NMM AST 0227 (Troschel the Younger).

25 See *Appendix 1*: MHSO 80277 (Heiden) and BM 1958,1006.2414 (Netherlands).

26 See *Appendix 1*: NMM AST0493. For more details on Johannes Buschmann, see the Webster Signatures Database, http://historydb.adlerplanetarium.org/signatures/ (accessed 13 April 2018).

27 For aspectaria on table clocks, see Jane Desborough, "The Changing Face of Time: The Making of the Modern Clock and Watch Dial 1550–1770", unpublished doctoral thesis, University of Leeds, 2016.

28 Silke Ackermann, "Astrological Scales on the National Maritime Museum Astrolabes", in Koenraad Van Cleempoel (ed.), *Astrolabes at Greenwich: A catalogue of the Astrolabes in the National Maritime Museum, Greenwich*, Oxford, Oxford University Press/National Maritime Museum, 2005, pp. 73–89. There is one Islamic astrolabe in the National Maritime Museum which shows the exaltations (AST0594) and a European version with both exaltations and depressions (AST0567).

FIGURE 6.2 Astrolabe, French, *c.* 1630; front, showing the Rojas projection, lunar volvelle and aspectarium (National Maritime Museum, Greenwich, no. AST0564).

days for medical treatment.[29] The third is a late fifteenth-century Italian astrolabe with astrological pointers for each planet.[30] Additional research is required on the significance and use of these instruments.

29 Astrological plate and lunar instrument, Germanisches Nationalmuseum, WI 1808. My thanks to Josefina Rodríguez-Arribas for alerting me to this object. See also her chapter in this volume, *infra*, pp. 46–50.

30 Astrolabe and astrological volvelle, Italian, late fifteenth century, Museum of the History of Science, Oxford, inv. no. 45127.

Finally, there is the unique item known as Mercator's astrological disc, now in the Historisches Museum, Basel. This is included in *Appendix 1* since the aspectarium appears in diagram form, unlike the text found on the astrolabes mentioned above, and can be compared to the other instruments. It consists of a wooden disc 375 mm in diameter, covered on each side with paper discs. The back has a set of instructions in Latin while the front has nine paper rules (one each for the seven planets and two for the lunar nodes), arranged in a wheel shape over a disc of astrological data. At the centre lies a smaller paper disc on which are printed seven straight lines, each labelled with the appropriate name and glyph of the aspects, radiating from a point on the circumference. On the left-hand side of the opposition line the aspects have been labelled with their Latin names and associated glyphs, while the equivalent lines on the right-hand side have been labelled with their respective harmonies, such as *dilectio perfecte* ("perfect love", i.e. harmonious) for the trine aspect and *discordia* ("discord", i.e. disharmonious) for the quartile aspect. This combination of lines, glyphs and text labels is unusual and has not been found on any other object so far.[31]

Variations in Aspectaria Design

Having reviewed the prevalence of aspectaria on instruments, it is useful to consider the range of designs used by different makers. As with any creative process, instrument makers were keen to demonstrate both technical prowess and artistic flair in their work, and taking a deeper look at the aspectaria reveals a number of trends. *Appendix 1* classifies each aspectarium according to overall shape (complete geometric shapes or just single rays for each aspect), how the aspects are labelled (with Gothic numerals, astrological glyphs or full text), and whether the aspectarium is functional or not, depending on the presence of a zodiac scale, which would allow one to plot the relative positions and influences of the Sun, Moon and planets. *Appendix 2* lists each instrument maker according to his favoured aspectarium design. These two lists provide some insight into how these trends varied over time, geographical area and across generations of the same family.

In the first instance, it seems that the complete geometric shape of the zodiac circle filled with a square, triangle, hexagon and associated glyphs, as depicted in Alcabitius' *Book of the Introduction to Astrology* (*Fig. 6.3*), was the

31 Steven Vanden Broecke, "Dee, Mercator and Louvain Instrument Making: An Undescribed Astrological Disc by Gerard Mercator (1551)", *Annals of Science* 58 (2001), pp. 219–240.

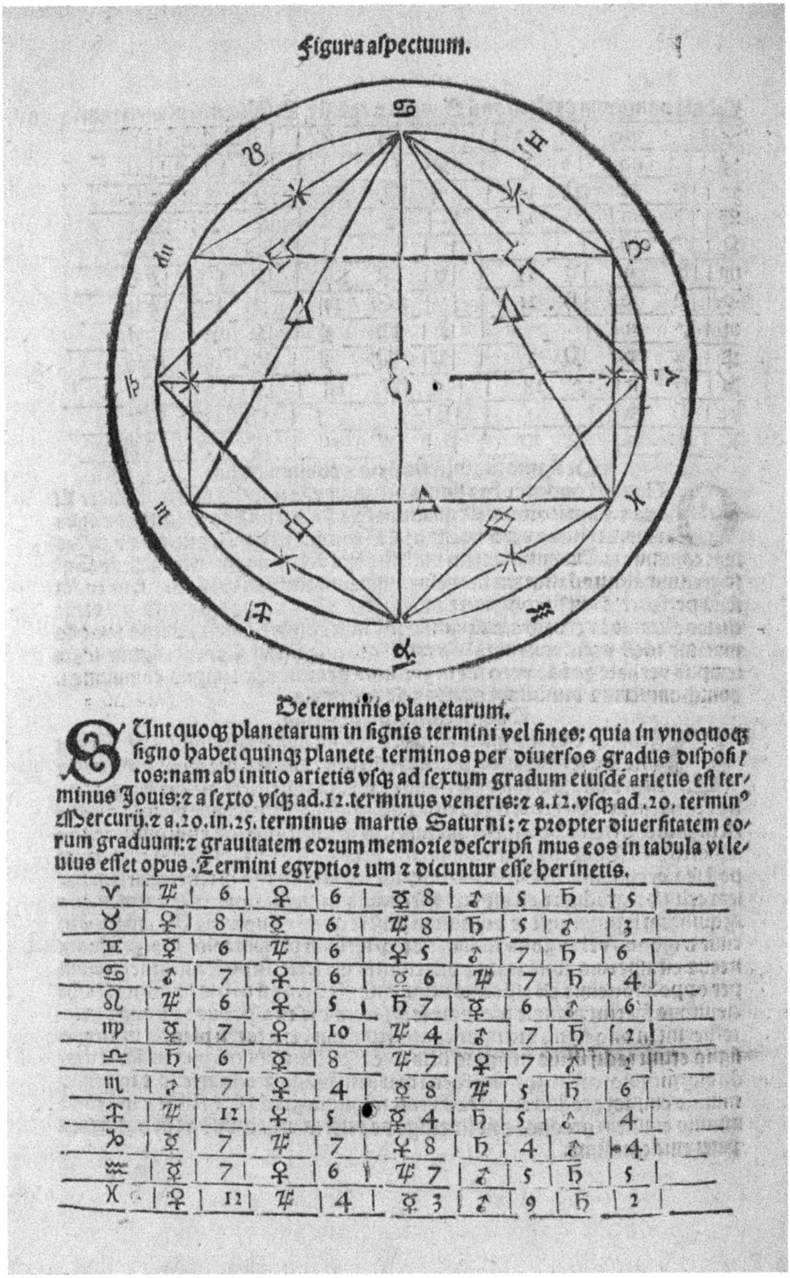

Figura aspectuum.

De terminis planetarum.

Sunt quoqꝫ planetarum in signis termini vel fines: quia in vnoquoqꝫ signo habet quinqꝫ planete terminos per diuersos gradus dispositos: nam ab initio arietis vsqꝫ ad sextum gradum eiusdē arietis est terminus Jouis: ꝛ a sexto vsqꝫ ad.12.terminus veneris: ꝛ a.12.vsqꝫ ad.20. termin⁹ Mercurij. ꝛ a.20.in.25. terminus martis Saturni: ꝛ propter diuersitatem eorum graduum: ꝛ grauitatem eorum memorie descripsimus eos in tabula vt leuius esset opus. Termini egyptiorum ꝛ dicuntur esse herinetis.

♈	♃	6	♀	6	☿	8	♂	5	♄	5
♉	♀	8	☿	6	♃	8	♄	5	♂	3
♊	☿	6	♃	6	♀	5	♂	7	♄	6
♋	♂	7	♀	6	☿	6	♃	7	♄	4
♌	♃	6	♀	5	♄	7	☿	6	♂	6
♍	☿	7	♀	10	♃	4	♂	7	♄	2
♎	♄	6	☿	8	♃	7	♀	7	♂	2
♏	♂	7	♀	4	☿	8	♃	5	♄	6
♐	♃	12	♀	5	☿	4	♄	5	♂	4
♑	☿	7	♃	7	♀	8	♄	4	♂	4
♒	☿	7	♀	6	♃	7	♂	5	♄	5
♓	♀	12	♃	4	☿	3	♂	9	♄	2

FIGURE 6.3 Astrological aspectarium; from *Alchabitius cum comento: Cum gratia et privilegio*, Venice, per Joan[n]em [et] Gregoriu[m] de Gregoriis fratres, 1502, fol. 3v.

prevailing design adopted by makers across all types of instrument during the entire period of the survey (1480–1650). An alternative was a ray diagram with a set of straight lines radiating either from the lunar index pointer or the aperture on the lunar volvelle which depicts the phase of the Moon. The earliest known occurrence of this design can be found on Mercator's astrological disc of 1551, described above. Here, the large size of the aspectarium volvelle gave Mercator the opportunity to label the rays with the full Latin names and harmonies of the aspects. He specifically mentions taking inspiration from astrological texts by Ptolemy, Alcabitius, John of Seville and Girolamo Cardano,[32] which resonates with the previous assumption about the origin of the complete diagram. For the remaining examples, the makers decided to compress the ray diagram into a smaller volvelle and the aspects were simply labelled with their respective glyphs. An exception to this is Hans Dorn (*fl.* 1490s), who preferred to label the aspects with gothic numerals with 3, 4 and 6 denoting respectively trine, square and sextile (*Fig. 6.4*).[33]

After Mercator, the ray diagram appears within two families of instrument makers in Nuremberg during the 1560s–1570s. While other Nuremberg makers chose the complete aspectarium, Christian Heiden and his son Johannes Paulinus (*fl.* 1570s) both adopted a ray diagram with full text labels. Meanwhile, Hans Troschel (*c.* 1549–1612) and his son (*c.* 1599–*c.* 1634) used the ray diagram style on their ivory diptych dials, although the aspects were simply labelled with the glyphs rather than the full text names. These two families were the exception: all other Nuremberg makers favoured the complete diagram seemingly based on the figure in Alcabitius' astrological textbook.

Meanwhile, across the English Channel, the situation was rather different. *Appendices 1* and *2* show that *all* London instrument makers used the ray diagram in their aspectaria, starting in the 1560s with Humfrey Cole, the first English-born instrument maker, continuing through to Elias Allen in the 1650s. This curious geographical concentration of one design may have its origins in Louvain and the work of Mercator who, as already seen, favoured this type of aspectarium. During the 1530s, Mercator privately taught several students in mathematics, cartography, astronomy and making scientific instruments. One such student was Thomas Gemini (*c.* 1510–1562), a Flemish engraver and printer who moved to London in the early 1540s. Over the next two decades, Gemini produced seven known instruments, including an astrolabe for King Edward VI, made around 1551, which contains much detailed astrological information. This instrument is believed to be modelled on Mercator's

32 *Ibid.*, p. 236.

33 *Appendix 1*: BM 1894, 0615.1.

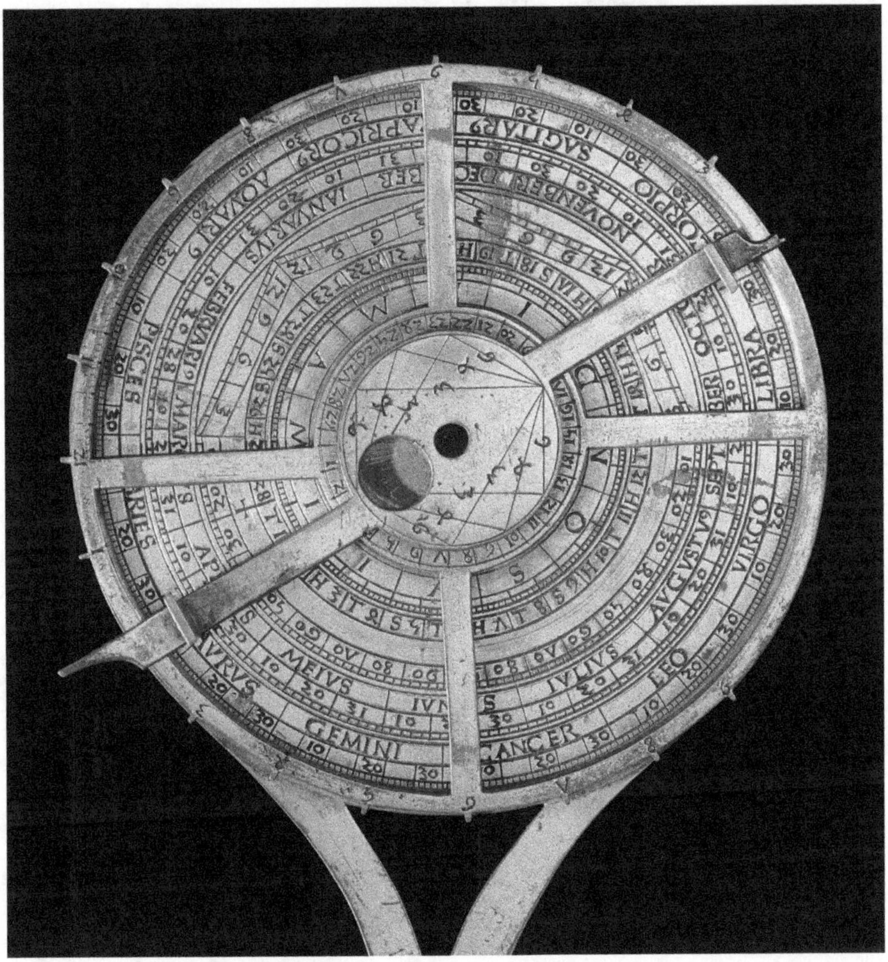

FIGURE 6.4 Silvered brass nocturnal, by Hans Dorn, Vienna, 1491. Note the aspects labelled with gothic numerals (British Museum, no. 1894,0615.1).

astrological disc,[34] with the astrological data most likely supplied by the English court philosopher, John Dee (1527–1609), who had worked with

34 Only the *mater* of the astrolabe now survives: NMM AST0567. See Gerard L'Estrange Turner, Koenraad Van Cleempoel, "A Tudor Astrolabe by Thomas Gemini and its Relationship to an Astrological Disc by Gerard Mercator of 1551", *The Antiquaries Journal* 81 (2001), pp. 400–409. For an overview of Gemini's career and a detailed description of each instrument produced see Turner, op. cit. (n. 1), pp. 12–20. For the King Edward VI astrolabe see *ibid.*, pp. 95–97, no. 1.

Mercator in 1548–1550.[35] The thread between Louvain and London continues with the work of Humfrey Cole who, like Mercator, incorporated the ray diagram into his instruments. Although no direct link is known between Gemini and Cole, there is a possible link through Dee, who taught navigation to the explorers in preparation for the 1576 voyage led by Martin Frobisher (1535?–1594) to find a north-west passage to Asia. The account books of one of the key investors in the voyage, Michael Lok (*c.* 1532–1620/22), show that "*li* 50 14 *s* 0 *d*" (£ 50 and 14 shillings) was paid to "Humfrye Cole and others" for instruments, books and maps. It is plausible that Cole and Dee would have known each other through this venture and could have exchanged ideas about instruments and astrology.[36]

After Cole's death in 1591, James Kynvyn and Charles Whitwell emerged as the next generation of London makers, both following Cole's choice of ray diagram for their aspectaria (*Fig. 6.5*). Whitwell also adopted Cole's innovative nocturnal design, whereby the user views the alignment between the north star (Polaris) and the nearby star Kochab through a slit, rather than using a central hole and index pointer, so there is a clear link here between Cole's work and Whitwell's later designs. Similarly, Whitwell's apprentice, Elias Allen, adopted the same aspectarium design. Thus one can see a distinctive lineage of designs being passed down successive generations of the London trade.

Examining *Appendices 1* and *2* in more detail also reveals how the full text names of the aspects were generally found on instruments where there was a large flat surface for inscriptions, such as metal nocturnals or the astrological disc by Mercator. Given that the diagram in Alcabitius' work (*Fig. 6.3*) does not include the full text names, it is interesting to consider other sources of inspiration. One possibility is the set of publications produced by the mathematician and cosmographer, Petrus Apianus (Peter Bienewitz, 1495–1552), who produced a guide to astronomy and various practical techniques in the *Cosmographia* (1524). Sixteen years later, Apianus was employed as Professor of Mathematics at the University of Ingolstadt and excelled himself by producing the *Astronomicum Caesareum* (1540), a richly coloured book filled with rotating paper volvelles superimposed onto astronomical diagrams. Written for Apianus' patrons, Charles V, and his brother, Ferdinand of Spain, the book

35 R. Julian Roberts, "Dee, John (1527–1609)", *Oxford Dictionary of National Biography*, Oxford University Press, online edition, May 2006, www.oxforddnb.com/view/article/7418 (accessed 15 November 2015).

36 James McDermott, "Humphrey Cole and the Frobisher Voyages", in Silke Ackermann (ed.), *Humphrey Cole: Mint, Measurement and Maps in Elizabethan England*, "British Museum Occasional Paper 126", London, British Museum Press, 1998, pp. 15–20: 16.

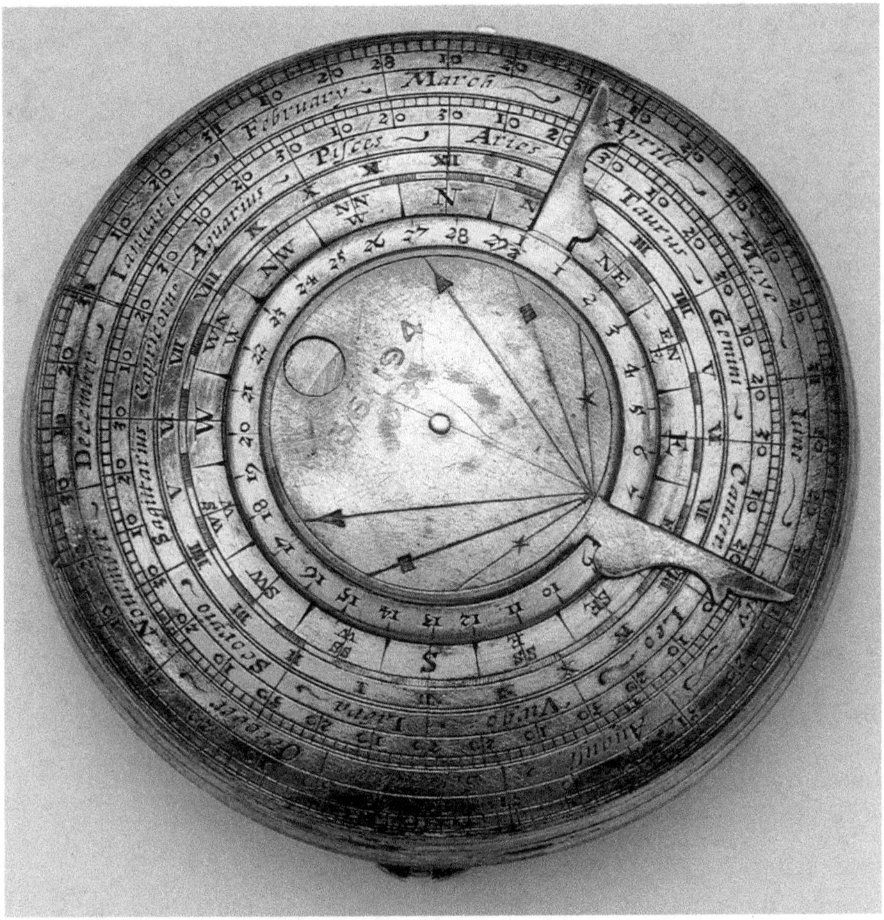

FIGURE 6.5 Ray diagram aspectarium on an astronomical compendium, by Charles Whitwell, about 1600 (National Maritime Museum, Greenwich, no. AST0468).
© NATIONAL MARITIME MUSEUM, GREENWICH, LONDON.

is composed of a series of propositions with which the reader is encouraged to perform calculations with the aid of the volvelles. The twenty-first proposition considers the planetary aspects and here one can see the full text names (*Fig. 6.6*),[37] as inscribed on some of the instruments listed in *Appendix 1*. Although the *Astronomicum Caesareum* was only produced in limited numbers

37 A full digital copy of the *Astronomicum Caesareum* can be viewed on the website of the Vienna University Observatory Rare Books Collection, www.univie.ac.at/hwastro/ (accessed 18 April 2018).

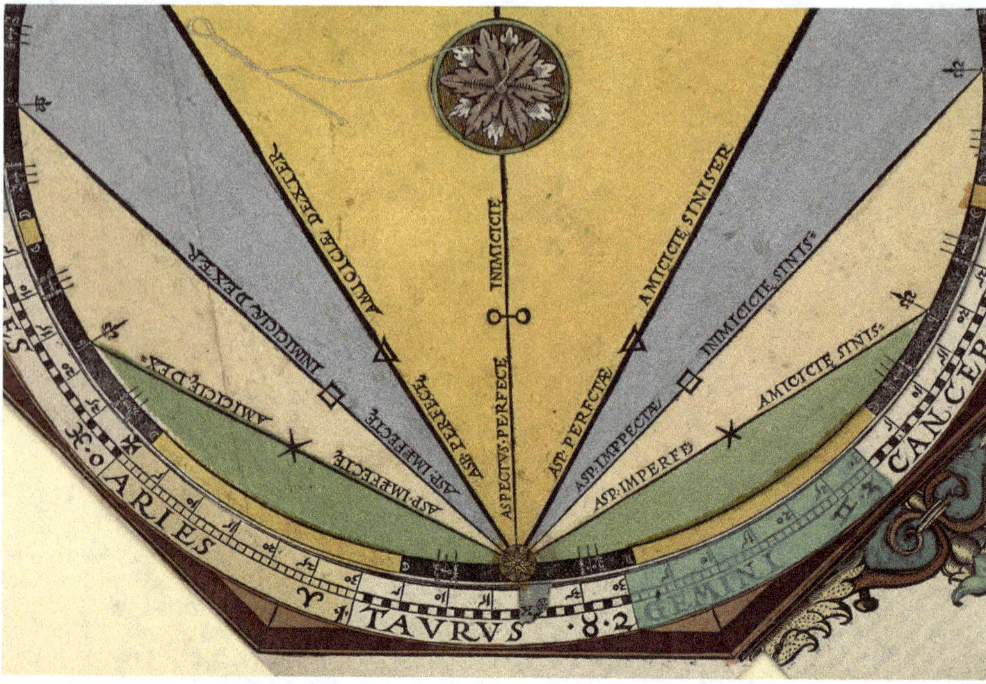

FIGURE 6.6 Paper volvelle to demonstrate the planetary aspects; from Petrus Apianus, *Astronomicum Caesareum*, Ingolstadt, 1540 (National Maritime Museum, Greenwich, no. ZBA7658).

compared to the many translations and 45 editions of the *Cosmographia*,[38] it seems reasonable to assume that the instrument makers of Nuremberg and other places would have been aware of its existence and may have drawn inspiration from this particular type of aspectarium.

Finally, let us consider whether the presence of an aspectarium on these instruments indicates whether they were used for astrological purposes or, as the current literature suggests, whether they offered a more aesthetic role as a

38 Over 110 copies are identified in Owen Gingerich, "A survey of Apian's *Astronomicum Caesareum*", in Karl Röttel (ed.), *Peter Apian. Astronomie, Kosmographie und Mathematik am Beginn der Neuzeit, mit Ausstellungskatalog*, second edition, Buxheim, Polygon-Verlag, 1997, pp. 113–122. See also Ronald Brashear, Owen Gingerich, Daniel Lewis, *Star Struck: One Thousand Years of the Art and Science of Astronomy*, San Marino CA, Huntington Library in association with the University of Washington Press, 2001, pp. 80–87.

decorative motif, an aide-mémoire of the aspects,[39] or perhaps as a memento mori highlighting the role of fate in one's life.[40]

In the first instance, one should consider the necessary elements required for an aspectarium to fulfil a functional role: a volvelle with the aspect lines; a zodiac circle to plot the positions of the planets using data from an almanac; and indexes for setting the positions of the Sun and Moon along the zodiac, if these particular aspects are desired. In his instructions for the astrological disc, Mercator offers this advice to would-be users:

> Next the beginner [in astrology] will wish to know the force of every [planet] with respect to the aspects, whether in relation to the other planets or to the other signs; let him turn the middle of the inner circle of the aspects to the index of this planet; at once the lines inscribed on this circle will show by what distance this planet looks upon all other planets and signs.[41]

A review of the makers' preferences in *Appendix 2* shows that it is the London instrument makers who included the necessary components, such as zodiac scales and index pointers, for the aspectaria to be functional. This is yet another indication of the likely link between Mercator's workshop in Louvain and the transfer of astrological knowledge through Thomas Gemini, John Dee and Humfrey Cole.

Glancing through *Appendix 1*, one can also see that the other instruments featuring functional aspectaria were metal nocturnals, particularly those produced by Ulrich Schniep in the 1580s (*Fig. 6.7*) and Caspar Vopel (1511–1561) in the 1540s and 1550s.[42] For these, the flat circular expanse of the nocturnal facilitated the inclusion of a zodiac scale and any additional pointers, particularly as the instrument had fewer components than a compendium and the maker could afford the time and spare materials to embellish the device with additional features. It is also curious to note that Vopel studied mathematics and medicine at the University of Cologne during the 1520s, which may suggest that he was more inclined to include a functional aspectarium, since astrology was an essential part of the medical curriculum, notably for determining

39 Hester Higton, "Glossary", in *Sundials at Greenwich: A Catalogue of the Sundials, Nocturnals and Horary Quadrants in the National Maritime Museum, Greenwich*, Oxford, Oxford University Press/National Maritime Museum, 2002, pp. 447–449.

40 For a discussion of sundials as memento mori and of sundial iconography more broadly, see Richard Dunn, "Sundials and the Arts", in *ibid.*, pp. 39–49: 46.

41 English translation from Vanden Broecke, op. cit. (n. 31), p. 239.

42 See EPACT, cit. (n. 19), for biographical details.

FIGURE 6.7 Nocturnal with aspectarium, zodiac scale and index pointers for the Sun and Moon, by Ulrich Schniep, Munich, 1581 (Museum of the History of Science, Oxford, inv. no. 51384).
© MUSEUM OF THE HISTORY OF SCIENCE, OXFORD.

propitious days for treatment. For example, students at the University of Vienna were encouraged to learn astrology through practical classes using paper astrolabes;[43] perhaps Vopel had a similar experience at Cologne which inspired his later instrument designs.

For Christoph Schissler and the majority of the Nuremberg makers, the lunar volvelle may have an aspectarium in the centre, but the surrounding numerical scales suggest that its primary function was to enable the owner to convert between lunar and solar hours when the dial was used as a Moon dial during the weeks before and after Full Moon. Similarly, one can see from *Appendix 1* that aspectaria are generally found on complex devices, such as compendia and table clocks, where timekeeping is the primary function and the aspectarium becomes a secondary feature, if space permits. For astronomical compendia, the listing reveals that the aspectarium is frequently found on the inner side of the outer leaf (Ib), an area which is not critical to the function of the inner dial or compass. Consequently, the maker has some choice in deciding what to include here, whether a list of cities and corresponding latitudes, perpetual calendars, saints' calendars or tide computer.[44] Equally, a maker may decide to embellish this area with a personalized feature relating to the intended owner, such as heraldic symbols or a pictorial design.[45] For the majority of makers, therefore, it seems that the aspectarium was indeed a decorative feature, which contributed towards the overall astronomical essence of the instrument but had little practical function and was only included according to the personal interests of the specific instrument maker or potential owner.

Conclusion

This review has shown that aspectaria were a popular feature on early modern astronomical instruments and were adopted by instrument makers in

43 Darin Hayton, "Instruments and Demonstrations in the Astrological Curriculum: Evidence from the University of Vienna, 1500–1530", *Studies in History and Philosophy of Science Part C: Studies in History and Philosophy of Biological and Biomedical Sciences* 41 (2010), pp. 125–134.

44 See Turner, op. cit. (n. 1), pp. 46–68 for a detailed description of the various components found in astronomical compendia.

45 For example, a pocket horizontal dial, by Augustine Ryther, 1585, MHSO, inv. no. 13314, has the inner side of the upper leaf personalized with the coat of arms and name of the owner, 'Geo: Seyntpoll', believed to be Sir George St Paul of Snarford and Carlton in Lincolnshire. See: www.mhs.ox.ac.uk/collections/imu-search-page/record-details/?TitInventoryNo=13314&querytype=field&thumbnails=on&irn=29571 (accessed 2 May 2018).

present-day Germany, France, Italy, England and the Netherlands. Aspectaria appeared on instruments produced during the period 1480–1650, and subsequent generations of makers within the same family adopted the design of their forebears. The most popular design favoured by instrument makers was the diagram included in the Latin translation of Alcabitius' *Book of the Introduction to the Craft of Astrology*, the predominant astrology textbook of the medieval period. This showed the four main aspects, represented as the complete geometric shapes of square, triangle and hexagon, labelled with the appropriate glyphs. For the majority of makers in mainland Europe, the aspectarium was a non-functional decorative motif included within the centre of a lunar volvelle, which was primarily intended to be used for converting between lunar and solar hours, rather than to determine planetary aspects.

By contrast, London instrument makers tended to favour a ray diagram aspectarium situated on a lunar volvelle within a zodiac circle, enabling users to set the volvelle for astrological calculations. As discussed, this localized pattern of similar design and function may be a legacy of early connections between the first generation of London's instrument makers, the philosopher John Dee and the Louvain-based maker Gerard Mercator, a craftsman whose surviving instruments demonstrate a strong interest in practical astrology.

Appendix 1: Aspectaria in Major Instrument Collections in the UK

The table below lists aspectaria found on early modern scientific instruments in four UK museum collections, plus two unusual examples located in Florence and one in Basel. The instruments are listed by type (astronomical compendium, diptych dial, etc.), arranged in chronological order within each category.

Columns 1–4: Type, maker, place and date

Details obtained from online museum catalogues and EPACT (www.mhs.ox.ac.uk/epact/).

Column 5: Location

Most aspectaria were found on the lunar volvelle, a brass disc inscribed with the age of the Moon from 1 to 29½ days, complete with an aperture which shows the accompanying phases in pictorial format. The leaf reference follows the general convention used for compendia and diptych dials; for example, Leaf IIb means the underneath side (b) of the second leaf (II) within a diptych dial.

Column 6: Shape

The overall shape of the aspectarium is described as one of two types:
Full: the most common form, using full geometric shapes for quartile (square), trine (triangle) and sextile (hexagon) (see *Fig. 6.3*);
Ray: a ray diagram, with straight lines radiating from the lunar aperture to indicate each aspect (see *Fig. 6.6*);
Incomplete: a diagram where the lines between the glyphs are either missing or incomplete.

Column 7: Labels

This column provides more detail on how the aspects are labelled:
Go: Gothic numerals (3, 4, 6);
Gl: glyphs (triangle, square and star symbols);
Te: Latin text (Trinus, Quadratus, Sextilis, Oppositio), sometimes combined with Greek (Trigonus, Tetragonus, Hexagonus).

Column 8: Functional

An aspectarium is classified as functional if it is embedded within a zodiac scale as a measure of the separation between Sun, Moon and/or various planets.

Column 9: Museum number

Holding institution and object number:
BM: British Museum, London (www.britishmuseum.org/research/collection_online/search.aspx);
HMB: Historisches Museum, Basel (www.hmb.ch/sammlung/uhren-und-wissenschaftliche-instrumente.html);
MHSO: Museum for the History of Science, Oxford (www.mhs.ox.ac.uk/collections/imu-search-page/);
NMM: National Maritime Museum, Greenwich (http://collections.rmg.co.uk/);
ScM: Science Museum, London (http://collectionsonline.nmsi.ac.uk/);
MG: Museo Galileo, Florence (http://catalogue.museogalileo.it/).

Type	Maker	Place	Date
Astrolabe	Unsigned	France	*c.* 1630
Astrological disc	Mercator, Gerard	Louvain	1551
Astronomical compendium	Dorn, Hans (attr.)	Hungary	1481
Astronomical compendium	Unsigned	Germany	1550–1600
Astronomical compendium	Schissler, Christoph	Augsburg	1550–1600
Astronomical compendium	Schniep, Ulrich	Munich	1553
Astronomical compendium	Klieber, Ulrich	Augsburg	1555
Astronomical compendium	Schissler, Christoph	Augsburg	1557
Astronomical compendium	Schissler, Christoph	Augsburg	1560
Astronomical compendium	Schissler, Christoph	Augsburg	1563
Astronomical compendium	Schissler, Christoph (attr.)	Augsburg	1565–1575
Astronomical compendium	Schissler, Christoph	Augsburg	1566
Astronomical compendium	Schniep, Ulrich	Munich	1566
Astronomical compendium	Menusier, Hierosme Vize	Pau	1567
Astronomical compendium	Reichle, Melchior	Germany	1568
Astronomical compendium	Cole, Humfrey	London	1569
Astronomical compendium	Schniep, Ulrich	Munich	1570
Astronomical compendium	Heiden, Joannes Paulinus	Nuremberg	1577
Astronomical compendium	Schissler, Christoph	Augsburg	1588
Astronomical compendium	Kynvyn, James	London	1593
Astronomical compendium	Klieber, Tobias	Augsburg	1595
Astronomical compendium	Linden, Johann Anton	Heilbronn	1596
Astronomical compendium	Unsigned	Italy	1595–1605
Astronomical compendium	Whitwell, Charles	London	1600
Astronomical compendium	Allen, Elias	London	1600–1650
Astronomical compendium	Unsigned	Germany	1600–1625
Astronomical compendium	Schissler, Christoph	Augsburg	1608
Astronomical compendium	Unsigned	England	1900–1950 (fake)
Astronomical table clock	Rauber, Caspar (attr.)	Germany	*c.* 1575

Location	Shape	Labels	Functional	Museum number
Middle volvelle on front	Full	Gl	No	NMM AST0564
Central volvelle and text	Ray	Gl + Te	Yes	HMB 1876.20
Lunar volvelle Leaf Ia	Full	Go	Yes	MHSO 43055
Lunar volvelle Leaf Ib	Full	Gl	No	NMM AST0466
Lunar volvelle Leaf Ib	Full	Gl	No	MHSO 55301
Lunar volvelle Leaf Ib	Full	Gl	No	MHSO 52296
Lunar volvelle Leaf Ib	Full	Gl	No	MHSO 54402
Lunar volvelle Leaf IIIa	Full	Gl	No	BM 1913, 0710.1
Lunar volvelle Leaf IIIb	Full	Gl	No	NMM AST0410
Lunar volvelle Leaf Ib	Full	Gl	No	BM 1867,0705.23
Lunar volvelle Leaf Ib	Full	Gl	No	BM 1855,1201.227
Lunar volvelle Leaf IIIa	Full	Gl	No	MHSO 35846
Lunar volvelle Leaf Ib	Full	Gl	No	MHSO 48219
Lunar volvelle Leaf Ia	Full	Gl	No	BM 1893,0616.7
Lunar volvelle Leaf Ib	Full	Gl	No	BM 1865,0620.5
Lunar volvelle Leaf Ib	Full	Gl	Yes	NMM AST0172
Lunar volvelle Leaf Ib	Full	Gl	No	MHSO 52029
Lunar volvelle Leaf IIb	Ray	Te	Yes	NMM AST0409
Lunar volvelle Leaf IIIb	Full	Gl	No	MHSO 49036
Lunar volvelle Leaf IIb	Ray	Gl	Yes	BM 1866,0221.1
Lunar volvelle on sliding plate	Full	Gl	No	BM 1867,0705.24
Reverse of leaf b, central section	Full	Go+Gl	No	BM 1857,1116.1
Lunar volvelle Leaf Ia	Full	Gl	Yes	BM 1855,1201.225
Lunar volvelle Leaf IIb	Ray	Gl	Yes	NMM AST0468
Lunar volvelle Leaf IIb	Ray	Gl	Yes	BM 1863,0929.2
Lunar volvelle Leaf Ib	Full	Gl	No	BM 1871,1115.9
Lunar volvelle Leaf Ia	Full	Gl	No	MHSO 49061
Lunar volvelle Leaf Ib	Full	Gl	Yes	NMM AST0219
Lunar volvelle at the centre of a planispheric astrolabe	Full	Gl	Yes	MG 3370

Table (*cont.*)

Type	Maker	Place	Date
Astronomical table clock	Weydmann, Lucas	Krakow	1648
Astronomical table clock	Unsigned	Augsburg	1680
Astronomical watch	Roberts, H.	London	1600–1610
Calendar watch	Jacobs, Coenrad	Leeuwarden	1610–1620
Calendar watch	Arthaud, Louis	Lyon	1645–1655
Diptych dial	Heiden, Christian	Nuremberg	1569
Diptych dial	Schissler, Hans Christoph junior	Augsburg	1582
Diptych dial	Unsigned	Netherlands	1595
Diptych dial	Troschel, Hans the Elder	Nuremberg	1607
Diptych dial	Trechsler, Christoph the Elder	Dresden	1622
Diptych dial	Troschel, Hans the Younger	Nuremberg	1624
Magnetic azimuth dial	Buschmann, Johannes	Augsburg	1600–1650
Mechanical equinoctial dial	Beyser, J.D.	Germany	1700–1800
Nocturnal	Unsigned	France	1500–1600
Nocturnal	Vopel, Caspar	Cologne	1543
Nocturnal	Vopel, Caspar	Cologne	1551
Nocturnal	Vopel, Caspar	Cologne	1557
Nocturnal	Cole, Humfrey	London	1570–1580
Nocturnal	Schniep, Ulrich	Munich	1581
Nocturnal	Leonae, A.	Rome	1589
Nocturnal	Unsigned	France	1597
Nocturnal	Pineau (attr.)	France	*c.* 1600
Nocturnal and sundial	Dorn, Hans	Vienna	1491
Perpetual calendar	Unsigned	England	1700–1800
Perpetual calendar	Unsigned	Germany	1700–1800
Square box horizontal dial	Unsigned	Italy	1500–1600
Square box horizontal dial	Schniep, Ulrich	Munich	1572

Location	Shape	Labels	Functional	Museum number
Lunar volvelle on upper dial	Full	Gl	No	BM 1867,0716.4
Lunar volvelle at the centre of a planispheric astrolabe	Full	Gl	Yes	BM 1958,1006.2129
Lunar volvelle Leaf IIa	Full	Gl	Yes	BM 1856,0429.1
Lunar volvelle within chapter ring	Full	Gl	No	BM 1888,1201.176
Lunar volvelle within calendar ring	Full	Gl	No	BM 1874,0718.66
Lunar volvelle Leaf Ia	Ray	Te	Yes	MHSO 80277
Lunar volvelle Leaf Ib	Full	Gl	No	NMM AST0462
Lunar volvelle Leaf Ib	Full	Gl	No	BM 1958,1006.2414
Lunar volvelle Leaf IIb	Ray	Gl	No	BM 1888, 1201.287
Lunar volvelle Leaf Ia	Full	Gl	No	MHSO 38997
Lunar volvelle Leaf IIb	Ray	Gl	No	NMM AST0227
Lunar volvelle inside lid	Full	Gl	No	NMM AST0493
Lunar volvelle inside lid of the case	Full	Gl	No	MHSO 41087
Lunar volvelle	Full	Gl	No	NMM AST0124
Lunar volvelle	Full	Gl+Te	Yes	ScM 1883–131
Lunar volvelle	Full	Gl+Te	Yes	BM 1895,0319.1
Lunar volvelle	Full	Gl+Te	Yes	MHSO 33871
Lunar volvelle	Ray	Gl	No	BM 1857,1116.2
Lunar volvelle	Full	Gl+Te	Yes	MHSO 51384
Lunar volvelle	Full	Gl	Yes	NMM AST0130
Lunar volvelle	Full	Gl	No	NMM AST0118
Lunar volvelle	Ray	None	Yes	MG 2502
Lunar volvelle	Incomplete	Go	Yes	BM 1894,0615.1
Lunar volvelle	Full	Gl	No	MHSO 78935
Lunar volvelle	Full	Gl	No	MHSO 53916
Lunar volvelle Leaf Ia	Incomplete	Gl	No	MHSO 51587
Lunar volvelle Leaf IIb	Full	Gl	No	MHSO 44205

Table (*cont.*)

Type	Maker	Place	Date
Square box horizontal dial	Volckmer, Tobias	Braunschweig	1584
Square box horizontal dial	Trechsler, Christoph the Elder	Dresden	1613
Universal equinoctial dial	Schissler, Christoph	Nuremberg	*c.* 1570
Vertical disc dial	Schniep, Alexius	Vienna	*c.* 1600

Appendix 2: Scientific Instrument Makers and their Preferred Choice of Aspectarium Design

This table details the preferred choice of aspectarium for each instrument maker featured in *Appendix 1*.

Columns 1–3: Maker, place, dates
Details of the instrument type with maker, origin and date information from the online museum catalogues or EPACT (www.mhs.ox.ac.uk/epact/).

Column 4: Shape
The overall shape of the aspectarium is described as one of two types:
Full: the most common form, using full geometric shapes for quartile (square), trine (triangle) and sextile (hexagon) (see *Fig. 6.3*);
Ray: a ray diagram, with straight lines radiating from the lunar aperture to indicate each aspect (see *Fig. 6.6*).

Column 5: Labels
This column provides more detail on how the aspects are labelled:
Go: Gothic numerals (3, 4, 6);
Gl: glyphs (triangle, square, and star symbols);
Te: Latin text (Trinus, Quadratus, Sextilis, Oppositio), sometimes combined with Greek (Trigonus, Tetragonus, Hexagonus).

Location	Shape	Labels	Functional	Museum number
Lunar volvelle Leaf IIIa	Full	Gl	Yes	MHSO 34940
Lunar volvelle Leaf Ia	Full	Gl	No	MHSO 49305
Lunar volvelle within chapter ring	Full	Gl	No	BM 1922,0705.3
Lunar volvelle	Full	Gl	No	MHSO 33775

Column 6: Functional
An aspectarium is classified as functional if it is embedded within a zodiac scale as a measure of the separation between Sun, Moon and/or various planets.

Maker	Location	Dates or period active	Shape	Labels	Functional
Allen, Elias	London	*c.* 1588–1653	Ray	Gl	Yes
Arthaud, Louis	Lyon	1612–1662	Full	Gl	No
Beyser, J.D.	Cologne	*fl.* 1700–1800	Full	Gl	No
Buschmann, Johannes	Augsburg	*c.* 1632–1676	Full	Gl	No
Cole, Humfrey	London	*fl.* 1568–1591	Full + Ray	Gl	Yes
Dorn, Hans	Vienna	*c.* 1435–1509	Full	Go	Yes
Heiden, Christian	Nuremberg	1526–1576	Ray	Te	Yes
Heiden, Joannes Paulinus	Nuremberg	*fl.* 1570s	Ray	Te	Yes
Jacobs, Coenrad	Leeuwarden	*fl.* 1610	Full	Gl	No
Klieber, Tobias	Augsburg	1545–1619	Full	Gl	No
Klieber, Ulrich	Augsburg	1529–1606	Full	Gl	No
Kynvyn, James	London	*fl.* 1590–1610	Ray	Gl	Yes
Leonae, A.	Rome	*fl.* 1580s	Full	Gl	Yes
Linden, Johann Anton	Heilbronn	*fl.* 1585–1605	Full	Gl+Te	No

Table (*cont.*)

Maker	Location	Dates or period active	Shape	Labels	Functional
Menusier, Hierosme Vize	Pau	*fl.* 1567–1586	Full	Gl	No
Mercator, Gerard	Louvain	1512–1594	Ray	Gl+Te	Yes
Pineau (attr.)	France	*fl.* 1600	Ray	-	Yes
Rauber, Caspar	Germany	*fl.* 1570s	Full	Gl	Yes
Reichle, Melchior	Germany	*fl.* 1568–1569	Full	Gl	No
Roberts, H.	London	*fl.* 1600–1610	Full	Gl	Yes
Schissler, Hans Christoph junior	Augsburg	*c.* 1561–*c.* 1626	Full	Gl	No
Schissler, Christoph	Augsburg	*c.* 1531–1608	Full	Gl	No
Schniep, Alexius	Vienna	*fl.* 1571–1613	Full	Gl	No
Schniep, Ulrich	Munich	*fl.* 1545–1588	Full	Gl	No
Trechsler, Christoph the Elder	Dresden	*fl.* 1572–1624	Full	Gl	No
Troschel, Hans the Elder	Nuremberg	*c.* 1549–1612	Ray	Gl	No
Troschel, Hans the Younger	Nuremberg	*c.* 1599–*c.*1634	Ray	Gl	No
Volckmer, Tobias	Braunschweig	*fl.* 1594–1624	Full	Gl	Yes
Vopel, Caspar	Cologne	1511–1561	Full	Gl+Te	Yes
Weydmann, Lucas	Krakow	*fl.* 1648	Full	Gl	No
Whitwell, Charles	London	*fl.* 1582–1611	Ray	Gl	Yes

CHAPTER 7

Italian Astronomical Clocks as Public Astrological Machines

Marisa Addomine

A personal interest in ancient monumental clocks and research into public astronomical timekeepers has led to a closer examination of the occurrence of aspectaria and zodiacs on a number of typically Italian clocks and of the methods used to obtain information from such dials. Although there has been little recent discussion of how to use these indicators, which have generally been considered as no more than curiosities or fashionable decorations, the widespread interest in astrology in the late medieval and Renaissance periods suggests that there is a more specific story to be told. This chapter considers, therefore, the occurrence of these distinctive dials and some sources that reveal their intended use.*

Astronomical Clocks in Public

European public astronomical clocks dating back to the Renaissance period are concentrated in two areas, Germany and German-speaking countries in the north, Italy in the south. When discussing monumental clocks, most horological scholars still rely on Alfred Ungerer's text, which takes into account only five examples in Italy, poorly described.[1] An article written sixty years later by Emmanuel Poulle was still based on Ungerer's reports as far as Italy is concerned.[2] More recent research, however, including archival descriptions of astronomical turret clocks and partial relics, has led to the identification of

* I am grateful to Marisa Cotti and Leontina Diemmi for their wonderful help in my first steps in astrology, to Chris McKay for his invaluable hints and suggestions and, last but not least, to my husband Daniele Pons, who shares in all my horological research.

1 Alfred Ungerer, *Les horloges astronomiques et monumentales les plus remarquables de l'Antiquité à nos jours*, Strasbourg, chez l'auteur, 1931, pp. 349–381.

2 Emmanuel Poulle, "Pour une typologie de l'horlogerie astronomique médiévale", *Comptes rendus des séances de l'Académie des Inscriptions et Belle-Lettres* 141, 2 (1997), pp. 635–662: 651–653.

 | DOI:10.1163/9789004381438_009

more than forty such timekeepers, with a higher frequency in north-eastern Italy but none south of Rome, other than some interesting examples in Sicily.[3] Many of these Italian astronomical clocks share a distinctive design, which differs from that of their German counterparts. The latter very often include an astrolabic dial, with an eccentric zodiac belt, whereas Italian makers used different mechanisms and representations of the celestial phenomena.

The first astronomical turret clock in Italy (*Fig. 7.1*) is believed to be the one designed by Jacopo Dondi Dall'Orologio (1293–1359), who taught medicine at the University of Padua and was well known for his skill in astronomical and horological matters. It was made for the city of Padua in 1344, following a request from Ubertino I de' Carraresi, lord of the city at that time. The original clock was destroyed in 1390 during a war between the Carraresi family and the Venetian-Milanese coalition, but the Padua historian Andrea Gloria reported in his notes on the history of the clock that in 1436 a younger relative of Jacopo Dondi, Novello Dondi Dall'Orologio, who had access to the original drawings, made a faithful replica, helped by two blacksmiths, Giovanni and Gian Pietro dalle Caldiere.[4] This was the first clock in the Italian style, with several concentric discs, the innermost carrying the astrological aspects (see below). The Padua clock has survived with the orginal configuration of its dials, although the mechanism has been heavily modified. It should therefore be considered the prototype for all Italian astronomical clocks: they share the same configuration of the dials, including examples in which the motions were obtained using different mechanical solutions. Such dials can be found on clocks in Brescia (Paolo Gennari, 1546), Clusone (Pietro Fanzago, 1583; *Fig. 7.2*), Arezzo (Felice da Fossato, 1552) and, with some variations, in Cremona (Francesco and Giovan Battista Divizioli, 1588) and Mantua (Bartolomeo Manfredi, 1473), to mention just the older examples. A famous example of a concentric disc astronomical dial in the Italian style can also be found outside Italy: Nicolaus Kratzer's and Nicolas Oursian's clock at Hampton Court in England, made in 1540.

3 Some preliminary results of the survey were presented in Marisa Addomine, "Italian Astronomical Clocks—A Quick Virtual Grand Tour", paper delivered at the British Horological Institute, 12 September 2012.

4 Andrea Gloria, "L'Orologio di Jacopo Dondi nella Piazza dei Signori di Padova, modello agli orologi più rinomati d'Europa", *Atti e Memorie della R. Accademia di Scienze Lettere ed Arti in Padova* 286 (1884–1885), pp. 233–293.

FIGURE 7.1 Astronomical turret clock in Padua, by Jacopo Dondi and Novello Dondi dall'Orologio. The original clock was made in 1344 and is considered the prototype for other Italian monumental astronomical/astrological clocks.

Aspectaria on Public Clocks

A Renaissance astronomical clock, showing the Sun's position in the zodiac, necessarily has a 24-hour dial, with the hour hand completing its revolution in a day, thus mimicking the Sun's position in the zodiacal belt. The zodiacal belt is usually drawn inside the ring carrying the numbers for the hours. The Moon window is also placed on a moving plate, so that its current position in the zodiac can be easily read. In addition, Renaissance astronomical clocks and watches often have on their main dial—the one housing the hour hand—a distinctive set of geometrical figures: a triangle, square and hexagon. They

FIGURE 7.2 Astronomical turret clock in Clusone, by Pietro Fanzago, 1583.

share one vertex, which coincides with the Moon phase indicator, usually a round aperture beneath which a disc carrying a smaller circle, either painted in gold or in white—against a dark background—passes, simulating the phase of the Moon (*Fig. 7.3*). These features are also called aspectaria, visual aids to help identify the relative angular positions of the Sun and Moon as seen from Earth and giving important astrological information.[5]

As noted above, the aperture (*oculus*) that shows the lunar phase coincides with the vertex shared by the geometric figures. Astrologers stated that the Sun was in trine aspect to the Moon if it was on one of the two other corners of the triangle. It was square if it was on one of the two corners of the square adjacent

5 See the chapter by Louise Devoy in this volume, *infra*, pp. 104–124.

FIGURE 7.3 Astronomical desk clock, built for Pope Pius V, by Giovanni Maria Barocci, *c.* 1570 (present location unknown). One of the dials shows the astrological aspects.

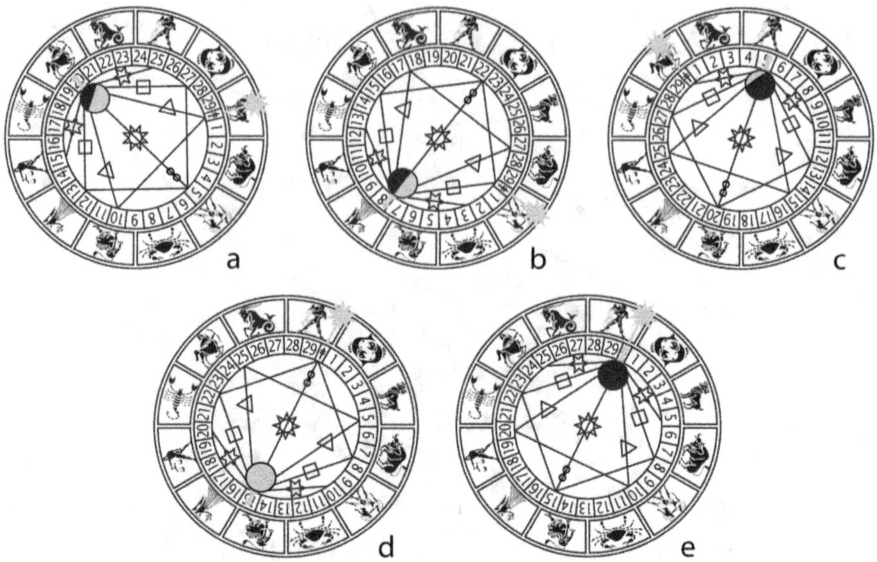

FIGURE 7.4 The aspects, the reciprocal angular positions of the Sun (drawn over the zodiacal belt) and Moon: a) trine; b) square; c) sextile; d) opposition; e) conjunction (drawing: Daniele Pons).

to the *oculus* and in sextile if on one of the two adjacent corners of the hexagon. The two other important configurations were opposition, when Sun and Moon were 180° apart, and conjunction, when they coincided (*Fig. 7.4*). Astrologically, the full power of an aspect was obtained when the angular relationsip was exact, but the few degrees either side of the vertices retained some power, the closer the stronger.

While these dials for the Moon phases are well known, most horological historians simply mention them in relation to defining the best times for agricultural activities—for instance, when to sow seed or plough—a practice still followed in some rural areas in Europe. Typically, no explanation is given for the geometric figures and their use. Poulle's analysis of the different types of astronomical clock dial includes, for example, a sentence worth quoting. Writing about the aspectaria in the middle of dials, he notes that:

> This separation [between Sun and Moon] is similarly represented by the geometric figure inscribed on the disc which turns with the pointer for the Moon, a figure reproducing the astrological terms for the separation between the two luminaries, so that astrologers talk of a square, trine or sextile aspect: such a figure is called astrological by commentators, *but*

> *there is no astrology in it*: the aspects are no more than a visual aid to show the current stage of the lunar month [...].[6]

While Poulle's extensive knowledge of medieval and Renaissance astronomical instruments and methods is undeniable, it seems unlikely that a clockmaker would have indulged in some mechanical complications just to include mementos. Aspectaria on timekeepers are too widely encountered to be relegated to the role of fashionable details. Furthermore, why should a Renaissance gentleman be interested in mere astronomical information? Watches and most domestic clocks, as well as turret clocks, were not sufficiently accurate to keep time to the standard required by astronomers. Astrological methods were more qualitative, however, and even the approximate information given by a common clock could be sufficiently meaningful in that context.

Astrological Readings

As a first attempt at tackling this issue, two acquaintances with long astrological experience were asked to read—based on their understanding—images of Italian dials, in an attempt to discern what an astrologer of the past might have read from specific configurations of the Sun and Moon indicators. Both responded that while they recognized the aspectaria patterns in the centre of the dials, modern astrology relies on reciprocal angular positions of *all* the planets: most Italian clocks had no planetary hands, and they could not make any astrological reading from lunar-solar aspects alone.

In order to understand the practical use of the dials, therefore, it is necessary to focus on the role of astrology in late medieval and Renaissance times. The major astrological sources were consulted to try to understand whether different methods might have been used and whether Sun and Moon alone could give meaningful information. Medieval and Renaissance astrological theory was based principally on Ptolemy's *Tetrabiblos*, which described in full the concept of planetary aspects, including the relationships between the positions of the Sun and Moon in the zodiac, which could be easily read on an

---

6 "Cet écart est également matérialisé par la figure géometrique inscrite sur le disque qui tourne avec l'aiguille de la lune, figure qui reproduit les formulations astrologiques de l'écart entre les deux luminaires dont on dit qu'ils sont en quadrature, en aspect trigone ou en aspect sextile; une telle figure est dite astrologique par les commentateurs, *mais il n'y a pas d'astrologie là-dedans*: les aspects ne sont qu'une façon de visualiser à quelle étape du mois lunaire on se trouve rendu [...]"; Poulle, op. cit. (n. 2), p. 652, my emphasis and translation.

Italian dial. It also described the aspects between the two bodies and the current Moon phase, which had further astrological significance. Each zodiac sign was assigned a set of qualities, including cold or warm, wet or dry, female or male, etc. They were also associated with one of the four elements (air, water, earth and fire) and ruled by one of the planets. Each of these properties was important for the casting and interpretation of a horoscope. The astrologer obtained additional information from the dynamic path of each celestial body with respect to the sign; for instance, the Moon just approaching a sign, at its centre, or leaving it led to different interpretations.

The Mantua clock seems to be the only example showing planetary hours, according to which each hour is linked to a planet (Sun and Moon being considered planets). Planetary hours were crucial for deciding which activities might be successful at any particular time of day and which were to be avoided. Some interesting examples of these relationships and the daily situations in which they might be applied are discussed below.[7] It is also worth noting that Sun and Moon positions, as represented on a 24-hour dial, reflect their actual place in the sky from a local terrestrial point of view, so they are real astronomical data, obtained in accordance with a Ptolemaic model, whereas planetary hours have no observational or astronomical correspondence and must be considered as purely astrological information.

Eclipses were another potentially significant influence, although the only Italian Renaissance turret clock that might show them—thanks to a dragon-shaped hand for *Caput Draconis* and *Cauda Draconis* (the head and tail of the dragon, indicating the lunar nodes)—is the one in Cremona, made by the Diviziolis, father and son.

The possible combinations of all these indicators gave a huge range of elements from which the astrologer might choose, for example in making health-care determinations. Indeed, medicine was one of the most important contexts for the application of astrology. Determining the best moment to undergo a surgical intervention, start a therapy, or even times at which to avoid these procedures was common practice, with medical practitioners expected to be adept at such astrological interpretation. The medical astrologer would, however, also draw on his own medical experience and emphasize or downplay those astrological factors which seemed more or less in tune with his preferred diagnosis or therapeutic approach.

Medieval medicine was largely based on Greek and Latin texts which had reached Europe thanks to Islamic cultural mediation. One of the seminal texts

7 Planetary hours also appear on the astronomical clock in St Paul's cathedral in Münster, Germany, made in 1542.

was Galen's *De Diebus Decretoriis*. This was translated into Latin in the second half of the thirteenth century, probably by William of Moerbeke,[8] and soon spread across Europe, although some of Galen's writings offering astrological hints to medical practitioners were already circulating in Europe by the ninth century.[9] Islamic medical astrology also drew on Galen's writings, contributing further to the long survival of his theories. *De Diebus Decretoriis* appears to have been used as a textbook in universities and was on medical scholars' bookshelves in Padua too.[10] Astrological hints intended for use by physicians appear in Book III.[11] On careful reading, it is notable that although Galen states that a full planetary diagram could provide more detailed information, he remarks that for most purposes the aspects of the Sun and Moon were to be considered sufficient for obtaining the essential information needed to offer medical advice to the patient. So-called Vitruvian clocks, water-driven, could bear both Sun and Moon indicators, so they could have been used as a help even in Galen's time. No such clock survives and our main source, referring to a public clock and not to a planetary instrument, is in Cassiodorus' *Variae*.[12] Even without a clock dial to look at, the information about the reciprocal Sun and Moon aspect could be easily calculated by astronomical tables and perhaps used by medical operators for therapeutic purposes.

Mechanical Aids for Astrology

There was some precedent for the use of mechanical aids in astrology. Most famously, Giovanni Dondi's *astrarium*, made between 1348 and 1364 and presented in 1381 to Gian Galeazzo Visconti, Duke of Milan, was primarily intended

---

8 An interesting review on the subject, including Galen's works, can be found in Cornelius O'Boyle, *Medieval Prognosis and Astrology: A Working Edition of the* Aggregationes de crisi et creticis diebus; *With Introduction and English Summary*, Cambridge, Wellcome Unit for the History of Medicine, 1991, pp. 1–24.

9 W.R. Newman, Anthony Grafton (eds.), *Secrets of Nature: Astrology and Alchemy in Early Modern Europe*, Cambridge MA, MIT Press, 2001, p. 122.

10 Pietro d'Abano, a friend to Jacopo Dondi and a professor in Padua university, mentions Galen and his analysis of the Moon positions for therapy in *Conciliator controversiarum, quae inter philosophos et medicos versantur*, Venice, Giunti, 1565, "Differentia 104", pp. 154v–155r.

11 Claudius Galenus (Galen), *De Diebus Decretoriis Libri III*, Lyon, Gulielmum Rovillium, 1560, pp. 389–524.

12 M. Shane Bjornlie, *Politics and Tradition Between Rome, Ravenna and Constantinople: A Study of Cassiodorus and the Variae, 527–554*, Cambridge, Cambridge University Press, 2012, pp. 163–184.

as an astrological machine. It was used by the astrologers at court, especially during the subsequent Sforza period, when astrological indications became almost an obsession and a political matter.[13] The power of astrologers in many Italian courts can hardly be overestimated and a mechanical instrument, providing immediate visual information about the sky's configuration, irrespective of the time of day and weather, would have been an invaluable tool in their hands. An astronomical clock allowed a much easier and faster reading, avoiding the long and complex task of calculations based on astronomical tables, more accurate from a mathematical point of view, but whose use was labour intensive, time-consuming and prone to human error. Simply looking at the dial of a public astronomical clock, the apparent reciprocal positions of Sun and Moon could be read at a glance, helped by the aspectaria.

The Mantua Clock and Its Uses

But what of people outside the court? If only a professional astrologer could understand aspectaria, why did they appear on public clocks? Were common people, or at least some of them, able to obtain astrological information from these clocks? A first hint comes from the comparison of dials: the dial of the clock in Mantua, although modified throughout the centuries, showed seasonal hours that were later interpreted as planetary hours; in other words, purely astrological information (*Fig. 7.5*).

The city of Mantua was extremely rich and powerful during the Renaissance. Its rulers, the Gonzaga family, established strong trading and political networks, as well as links with the most important courts of Europe, through marriage and other strategies. As at other European courts, astrology was an instrument of political power. Luckily, some interesting documents about the Mantua clock survive. The more recent one, "Dell'eccellenza, parti e uso de l'horologio di Mantova" (On the Excellence, Components and Use of the Mantua Clock) by Francesco Philopono, dates to the end of the sixteenth century and contains a detailed technical description, including mechanical drawings, of the clock.[14] It was written about a hundred years after the clock's completion as a report to the Duke about some recent mechanical problems. It was also a tool

13 Monica Azzolini, *The Duke and the Stars: Astrology and Politics in Renaissance Milan*, Cambridge MA, Harvard University Press, 2013, passim.

14 Francesco Philopono, "Dell'eccellenza, parti e uso dell'horologio di Mantova", 1561: Mantova, Biblioteca Comunale Teresiana, Ms 999. Almost nothing is known about the author, indicated as Filopono in other sources. The name itself may be a pseudonym.

FIGURE 7.5 The astronomical/astrological dial of the Mantua clock, following reconstruction in 1992.

of self-promotion for the author, who was later assigned the task of restoring the clock.

The long text includes a description of the dials and gives further confirmation of the purpose of the unusual white and black curved sectors that are on the dial, relating them to planetary hours.[15] It also gives a short summary of the

15 "First, outside the dial, [the clock] shows the common hours / Second, the motion of the Sun along the zodiac signs and its degrees / Third, the motion of the Sun between the tropics of Cancer and Capricorn / Fourth, the motion of the Moon along the zodiac signs and its degrees / Fifth, the length of daylight, depending on the Sun / Sixth, the aspects of the Moon in relation to the Sun / Seventh, the days elapsed since the conjunction of the Moon / Eighth, the length and brevity of nights and days / Ninth, the planetary hours, by day and night / Tenth, the ascending zodiacal sign and degree / Eleventh, the ascending,

astrological data that can be obtained from the clock. Philopono's description of the mechanical structure is sometimes fuzzy and he gives incorrect information about the history and age of the clock. Nevertheless, he confirms the regular use of the dial for astrological purposes and the importance of this for the city community as well as for its rulers. Of even greater value, however, is an older source, a handbook written to give the reader precise directions for deriving astrological information from the clock's dial.

Before analysing that text, it would be useful to make a few remarks about the maker of the Mantua clock, Bartolomeo Manfredi, also known in contemporary sources as Bartolomeo Arlojo (Bartholomew of the Clock). The son of a clockmaker, he learned his father's craft but had no regular schooling, although contemporary chroniclers describe him as a clever and faithful consultant to the Duke, especially in astrological matters. At this time, an influential member of the Duke's court was Vittorino de' Rambaldoni, born to an aristocratic family originally from Feltre and usually known as Vittorino da Feltre (1378–1446). Vittorino was a humanist, educated at the University of Padua, and developed a new pedagogic method for the Duke's sons. His school, called Ca' Zoiosa (The Joyful House) was based on the assumption that children should enjoy what they studied, rather than having to endure school as a burden. His approach included field trips, the active involvement of students in the learning process and individual study plans according to each student's needs and attitudes. His school soon became famous and is considered one of the milestones in the history of pedagogy. Vittorino was also famous for his broad interests, including excellent astrological skills, and Bartolomeo Manfredi learned astronomy and astrology from him. Astronomical knowledge was obviously necessary, since the calculation of the clock's gear trains, which had to simulate the motions of the celestial bodies, could only be carried out with a sound knowledge of the subject.

In 1474, a Mantuan gentleman, named Pietro Adamo de' Micheli, invited some German typographers, including two brothers, George and Paul

Medium Coeli, descending and Imum Coeli degrees of the sky / Twelfth, the ringing of [religious] offices, called the 'Officialia' [Primo egli dimostra fuori dell'Ostensorio le hore comuni / Secondo il moto del Sole nelli segni e gradi / Tercio il moto del Sole tra li tropici del Cancro e del Capricorno / Quarto, il moto della Luna nelli segni e gradi / Quinto la quantità de li giorni, congiunta con il Sole / Sesto li aspetti che ha la Luna con il Sole / Settimo li giorni che sono dopo congiontio con la Luna / Ottavo, la longhezza e brevità delle notti e giorni / Nono le hore planetarie, diurne e notturne / Decimo, il segno e grado che ascende / Undecimo, gli angoli del Cielo che ascendono, mediano, descendono e della meggianotte / Duodecimo il sonar gli officii della città detta l'Officiala]"; *ibid.*, fol. 8r. Philopono also gives a long description of the history of planetary hours and of their presence on the Mantua clock; *ibid.*, fols. 56v–60v.

Butzbach, to settle in Mantua, offering his financial help and starting the first book printing in the city. He also wrote a book in three parts, *Della Dichiaratione de l'Horologio di Mantova* (On the Explanation of the Mantua Clock), and had it printed, probably by the Butzbach brothers. Its subject was the new public clock, installed one year earlier. The Italian term "dichiarazione" is presumably intended as "the sum of things the clock tells us about", so that it *speaks in clear terms* (from the Latin *declarare*) to the audience. One presumes that if the clock were a private astrological instrument, for the Duke's personal use alone, de' Micheli would have written a manuscript: choosing the newest medium, the printed book, he intended this knowledge to be available in several copies to be disseminated to a number of readers.

The existence of a first part of the book can be inferred from references in the second part, but it has been lost, as no library or archive appears to have a copy. The references suggest that the first part dealt with the fundamentals of astronomy. The second and third parts survive, however, in a few copies. In over 70 pages, in often almost vernacular late-fifteenth century Lombard prose, these describe the clock and its use with practical examples. They include astrology, although the author expects the reader also to draw from the standard astrological treatises to understand the full repertoire of possible combinations and interpretations.

The first pages of the second section introduce the subject through a typically Renaissance literary rethorical device. The author dreams of walking in a garden, in which a group of beautiful women—the personifications of Lombardy's most important cities—are enjoying a peaceful life. The most beautiful of them, obviously, is Mantua. Her deeply enamoured partner is a nobleman, an allegory for the Duke Ludovico Gonzaga (1412–1478). The lady holds a wonderful mirror, which is the new clock, and with her help the author can now describe its wonders and the information its dial reveals.[16]

After a few pages, there follows a short dissertation of the reciprocal roles of astronomy, astrology and mathematics. The author writes that,

> the design and construction of this clock called for the knowledge and intelligence of these four sciences [astrology, geometry, arithmetics and music], which are called mathematics: astrology is the lady, the other

16 Pietro Adamo de' Micheli, *Della dichiaratione de l'horologio di Mantova*, Mantova, [Butzbach Brothers], 1474; Florence, Biblioteca Nazionale Centrale di Firenze (BNCF), inc. E.6.4.21, fols. 76r–115v: fol. 78v. The text is fully transcribed in Rodolfo Signorini, *Il mirabile orologio di Mantova e diciannove disegni inediti del 1706*, Mantova, Editoriale Sometti, 2011, pp. 99–124.

> three are the maidservants. As the lady stays in her room and the servants in the kitchen, so astrology stays in the sky and the other three do not leave the ground [...].[17]

Astronomy and its mechanical counterpart, horology, were clearly intended to be instrumental to the essential knowledge of astrology, the art which allowed, or at least helped, man to take control of his own destiny.

After a short introduction, de' Micheli lists eight different readings that can be obtained from the clock:

1. the so-called common people's time, i.e. how many hours have passed since sunset;
2. which zodiac sign and degree the Sun is currently in;
3. the same for the Moon, and how to read its phase and the day of the lunar month;
4. the so-called four corners of the sky: ascendant, descendant, Medium Coeli and Imum Coeli;
5. the planetary hours, i.e. which planet rules the current hour;
6. the usual bell ringings for Mantua (bells were rung in different ways at noon and midnight and at other times during the day, which varied according to the seasons);
7. the astronomical (defined in the text as astrological) time, i.e. how many hours have passed since noon, according to the German style;
8. the length of day and night throughout the year.[18]

---

17 "[...] alla compositione et fabrica di questo horologio gli è stata necessaria la cognitione et intelligentia di queste quatro scientie: le quali si chiamano le mathematice: L'astrologia è la madonna, le altre tre son le serve: perciò che come la madonna sta in camera et le serve in cucina: così l'astrologia sta ne' cieli: le altre tre non si partono da terra [...]"; de' Micheli, op. cit. (n. 16), fols. 82v.

18 "This instrument performs eight functions or demonstrations. The first is to know the time in the common people's way; that is, how many hours have passed since sunset, according to the widespread and antique use of all Italian clocks. The second is to know and always see in which [zodiacal] sign and degree is the Sun. The third is to know and always see, in a similar manner, in which sign and degree is the Moon, how old it is, and its visible portion; that is, if it is either round, or half full, or empty [i.e. new Moon], and on which side, on the east or the west, and when it appears, and when it turns; that is, when it is in conjunction and when in opposition and when at quadrature, or half-quadrature [i.e. octile], or at sextile, or trine with respect to the Sun, on the right or on the left. The fourth function is to know the four corners of the sky; that is, which sign and degree is ascending at the east, which is at Medium Coeli, which is at west, and which is at Imum Coeli, that is, below the Earth. The fifth function is to know the hours of the seven planets; that is, which planet rules at any hour, because now rules one planet, then another. This function is extremely useful, because very often you may find it is written that if you want to ask

Several pages also contain precise instructions for the reading and astrological use of the ascendant, descendant, Medium Coeli and Imum Coeli.[19]

The third section contains information of particular significance. After a dissertation on the zodiac signs and the importance of astrology in everyday life, de' Micheli describes the virtues and negative influences of the different signs and of the Sun and Moon when in those signs; for example, when the Moon is in Taurus:

---

for a favour from a high priest or a nobleman, you must ask for it in the hours of Jupiter or the Sun. And if you want to be successfully heard by women, you must speak to them in the hour of Venus. You must reject with horror the hours of Saturn and Mars for any affair, as I will more extensively explain in the last part of my booklet. The sixth function is to know the hours in the special way of Mantuan citizens, that is the morning bell, the morning religious services, third, ninth, the midday services and those of midnight; that is, at which hour each of these bells rings, and this is useful to craftsmen, citizens, priests, courtesans and any other person. The seventh function is to know the astrological [i.e. astronomical] hours; that is, how many hours have passed since noon, according to the method which is currently used in Germany; there, when it is exactly noon, they ring 24 hours. The eighth and last function is always to know how long are the day and the night on any day. [Otto son gli effetti et demonstration di questo istrumento. El primo è di sapere quante hore sono secundo el vulgo: cioè quante hore son passate doppo el tramontar del Sole: secundo la comune et antiqua consuetudine de tutti li horologii de Italia. El secundo serà di sapere anci vedere continuamente in qual signo et in qual grado serà el Sole. El terzo serà di sapere et vedere similmente in qual signo et grado serà continuamente la Luna et quanti zorni haverà: et quanta se vederà: cioè se la serà rotunda o meza piena o vota et in qual parte serà, cioè in oriente o in occidente: et quando la se farà: et quando la darà volta: cioè quando la serà in coniuntione et quando in oppositione: et quando serà in quadratura: o meza quadratura: o in aspetto sextile: o trigono col Sole dextro o sinistro. El quarto effetto serà di sapere li quatro anguli del cielo: cioè qual signo et qual grado serà in ascendente over in oriente: qual serà in mezo cielo: qual serà in occidente: et qual serà in meza notte, over angulo de la Terra. El quinto effetto è di sapere le hore di sette pianeti: cioè qual pianeta regna qualunque hora perché sempre regna mo un pianeta mo un altro. Et di questo si piglia utilità grandissima: perché molte fiate troverai scritto che se tu vogli domandar gratia ad qualche prelato o signore: tu la déi domandare ad hora de Iove o del Sole. Et se tu vogli havere grata audientia da donne: tu gli déi parlare ad hora di Venere: et schiffare le hore di Saturno e di Marte in ogni tua facenda come più diffusamente nel ultima parte del mio libretto te dirò. El sexto effetto è di sapere le hore particulari de mantuani cioè: La campana dal dì: L'offitiala da matina: Terza: Nona: Offitiala da mezo dì et meza notte, cioè ad qual hora sona ciascuna de loro. Et di questo ne pigliaranno utilità Artifici: Citadini: Offitiali: Cortegiani: et ciascuna altra persona. El septimo effetto è di sapere le hore di astrologi cioè quante hore siano passate doppo el mezo dì: secundo el qual modo ozidì se usa per tutta la Alemagna: dove quando egli è precisamente mezo dì alora sonano vintiquatro hore. L'octavo et ultimo effetto è di sapere continuamente quanto sia longo el dì et la note da ogni tempo]"; *ibid.*, fols. 85r–86r.

19 *Ibid.*, fols. 88r–90r.

> Taurus is a dry and southern sign: stiff and dry, melancholic, earthly and feral. Its influences are on the neck and throat and on their illnesses. Its ruling planet is Venus. When the Moon is in Taurus, it's good to sow seed, to plant orchards, vines, [and] plant trees so that they can grow and last long. It is a good time to build, to take a wife and in general to do whatever you want to be stable and enduring. It is unfortunate for curing one's neck and throat, to begin a battle, to face the enemy; in general, to do those things that you wish to be finished quickly. The weather is cold and dry; the winds will be low, blowing at ground level. Clouds are dry, bringing no rain. If a man is born with Taurus ascendant he has a large face and sometimes has a hump [on his back]. If this does not happen naturally, sometimes it happens by accident during his life. He has a long nose and sad and black eyes. His hair is curly, his neck is larger and fatter than usual; he is timid; when he walks he looks at the ground. He is honest; quite faltering in his steps. Do not rely on him to direct women.[20]

The use of the aspects is also explained with examples. The interpretation of a trine or sextile configuration, for example, is as follows:

> When the Moon is in a sextile or trine aspect to the Sun, that day is good to talk and work with noblemen and princes; to ask for graces, offices, dignities; to ride horses; to begin jobs. In general, it is good for all those things which are related to noblemen, barons and princes, and if you have to deal with them. Note that if a baby is born with the Sun exactly in midheaven in a fire sign and in a sextile aspect to the Moon, that baby will attain great power and status. For a baby born in the nineteenth

20 "Tauro è signo fixo: meridionale: frigido e sicco: melencolico: terreo et ferino: et ha dominatione sopra el collo e la gola et sopra le infirmità de quelli. Et il pianeta signor de Tauro è Venere. Quando la Luna è in Tauro bono è seminare: piantar orti: far le vigne: inserire: piantar arbori a ciò che crescano presto et durino assai. Bono è edificare: menar moglie: et generalmente far ogni cosa che tu desideri sia stabile et duri assai. Malo è medicar el collo e la gola: cominciar battaglia: andar in campo contra lo inimico: cominciar camino: et generalmente far ogni cosa che tu desideri habia presto fine. Et fa il tempo frigido e sicco et gli venti bassi e propinqui alla terra: et genera nuvole sicce che discorreno per l'aere e non piovono. Et l'homo chi nasce quando Tauro è in ascendente ha la faza grande et alle volte gobba se non per natura almanco per accidente in processo di tempo: el naso longo: gli occhii nigri e gravi: gli capilli aspri over ellevati: el collo grosso e grasso quasi più cha 'l debito: verecundo: et quando va guarda in terra: honesto nel andare instabile et quasi vagabundo: ma guarda non gli dare donne in governo." *Ibid.*, fols. 97r–97v.

> degree of Aries and the Moon in the third degree of Taurus, that baby will become a powerful lord.[21]

The central disc of the Mantua dial, which also contains a representation of the Roman goddess Latona, shows the reciprocal positions of the Sun and Moon, giving a sound basis for understanding the method for using aspects.

Another long section illustrates the concept and use of planetary hours. For example:

> In Venus's hour it is good to buy gems, pearl necklaces, jewels, belts, bags, nice textiles, furs, wool or linen drapes, blankets, canvases, small crates, boxes, silver or pewter vases, sugar, cloth, rope, thread and such things; to discuss weddings; to marry a virgin; to take a wife; to spend time with friends; to go to market; to talk to women; to talk about adultery and about both honest and illicit relationships; to sing and to stay up late, to organize banquets, begin dancing, draw blood, take medications, talk to noblemen in order to obtain favours, to paint images, to make schemes and plots related to love. And if the young lovers were to know the remarkable virtue of this hour, they would always diligently wait for it for their love affairs. It is a bad time to start a battle, to go into a field, to forge iron, to go to ovens and furnaces, to build walls or houses, to change furniture, to start studying or similar things.[22]

---

21 "Quando la Luna è in aspetto sextile o trigono dextro o sinistro col Sole: quel giorno bono è parlare et praticar cum signori baroni: e principi: domandar gratie: offitii: dignità: cavalcar cavalli: entrar in offitii: et generalmente bono è far tutte le cose che si apertengono a signori: baroni e principi: et si hanno ad praticar cum loro. Et nota che s'el nascesse uno essendo el Sol nel angulo del mezo ciel precise in signo igneo et in aspetto sextile de la Luna colui ascenderebbe ad grandissima potentia e condition. Et chi nascesse col Sole nel decimonono grado de Ariete e la Luna nel terzo de Tauro colui ascenderebbe ad grandissima signoria." *Ibid.*, fol. 105r.

22 "Nel hora de Venere: bono è comprare gemme: gerlande de perle: zoglielli: cinture: borse: binde: velli: drappi bianchi de lana o di lino: lenzuoli: tiele: cassette: scatole: vasi d'argento o di peltro: zucaro: canevazo: soghe: filo: et simil cose: parlar de noze: sposar virgine: menar moglie: far compagnie: andar in mercato: parlar cum gran donne: tractar de adulterii: fornicationi: amori: lenocinii: cantar: far matinate: ordinar convivii: cominciar balli: cavar sangue: pigliar medicina: parlar a signori per impetrar gratie: far imagini: caracteri: et coniurationi ad amorem. Et se li giovani inamorati sapessero in tutto la virtù di questa hora cum dilligentia sempre in ogni suo atto amoroso la observarebbono. Malo è cominciar bataglia: andar in campo: far focine: forni: fornaci: edificar muri o case: mutar massaritia: cominciar studio e simil cose." *Ibid.*, fols. 107v.

FIGURE 7.6 Pietro Adamo de' Micheli, *Della Dichiaratione dell'Horologio di Mantova*, second edition, Mantova, Giacomo Ruffinelli, 1547, frontispiece.

A final chapter, expunged from the 1547 second printed edition of the text,[23] deals with the possible use of the astrological indications of the clock for a more unusual purpose (*Fig. 7.6*). According to the detailed directions, in the event of a burglary, the clock could help identify a thief and his home, and

23 P.A. de' Micheli, *Della dichiaratione dell'horologio di Mantova*, Mantova, Giacomo Ruffinelli, 1547.

even locate the hidden loot. Thus the first lines of a long passage dealing with the identification of the burglar indicate that one should look for the sign that is the astrological descendant (also called the western corner) at the moment of the burglary:

> If Aries or Scorpio, ruled by Mars, is in the western corner, the thief has red, ruffled hair, yellow eyes; he has a horrible look. His face is round, brave; he has a mark or a spot on his foot.[24]

Using astrology for detective work might sound exotic to the modern reader but was widely known in Islamic countries and Europe in the Middle Ages. De' Micheli identifies his source for this practice as 'Haly'—Haly Albohazen, or Abū l-Ḥasan al-Shaybānī, the great eleventh-century astronomer and astrologer.[25] This reveals again the path by which some astrological doctrines reached western Europe. A Latin version of Haly's astrological compendium, *Praeclarissimus liber completus in iudiciis astrorum*, was well known and widely adopted in European universities from the thirteenth century.[26]

Last but not least, it is worth examining a possible chain for the transmission of astrological knowledge in the Mantuan cultural context, from academy to clockmaker's shop. It has already been noted that the clockmaker, Bartolomeo Manfredi, learned astronomy and astrology from Vittorino da Feltre, himself strongly influenced by an eminent scholar from the University of Padua, Biagio Pelacani, a pupil of Prosdocimo de' Beldomandi. Prosdocimo, in turn, was a pupil of Giovanni Dondi, whose father Jacopo's dangerous friendships included Pietro d'Abano (Petrus Aponensis), the philosopher and astrologer prosecuted for his unorthodox theories. Such pathways hint at the close connections between thinkers, natural philosophers and craftsmen in an age when new ideas concerning the relationship between humanity and the universe were being born.

---

24 "Se nel angulo ocidental è Aries o Scorpio che son signorezati da Marte: el ladro è rosso: colli capilli ruffi e occhii gialdi: aspetto horribile: faza rotunda: audace: et ha un signo over macula in un piede"; de' Micheli, op. cit. (n. 16), fol. 109v.

25 "Haly says that, depending on the Sun being in an eastern, western, southern or northern [zodiacal] sign, the thief's home is to the east, west, south or north of the place of the robbery [Dice Haly che secuno ch'el Sole è in signo orientale: occidentale: meridionale o setentrionale: cosi la casa del latro è verso oriente: occidente: mezo dì o setentrione: guardando dal loco dove è fatto el furto]"; *ibid.*, fol. 115r.

26 Haly Halbohazen, *Praeclarissimus in iudiciis astrorum liber Halbohazen Haly filii Abenragel*, Basel, Henrichi Petri, 1551.

Conclusion

This chapter has attempted to show that careful analysis of Mantuan sources sheds new light on the practical use of Italian astronomical clocks for astrological purposes. Above all, it has demonstrated that they were intended to be an important astrological tool for the whole, or at least most of, the community. Yet interest in aspectaria disappeared after the mid-seventeenth century as far as Italian public clocks are concerned. Indeed, while the extraordinary clock in Cremona originally had these distinctive geometrical patterns in its central disc, they are no longer visible today. They are well documented in old drawings and still visible in a photograph taken by Alinari Brothers in about 1940. As part of a restoration in 1978, however, the aspectaria were covered with a coat of black paint, a decision justified as erasing any trace of what was considered a past superstition!

CHAPTER 8

Astrological Time in Public Space: The Görlitz Arachne (1550) and Planetary Hours

Richard L. Kremer

For most of its adherents, the practice of early modern astrology occurred in private spaces.* The armamentarium required to generate and transport astrological knowledge—its theoretical texts and handbooks, astronomical and astrological tables, paper and brass instruments, globes and armillary spheres, printed broadside calendars and prognostications, horoscopes and charts—resided in small, portable artefacts. These material objects were stored, consulted and manipulated in private, in the interior spaces of the home, the palace, the monastery, the library, or the astrologer's consulting room. Images of the early modern astrologer at work, either alone or with a client, generally place him in these private contexts (*Fig. 8.1*). Although he might look out at the starry sky, the astrologer conducted his craft inside, secluded from public view, surrounded by books.[1]

Yet the astrological cosmos of early modern Europe derived its meaning from the positions of the celestial bodies, primarily the planets. For the astrologically inclined, the influences of those bodies on terrestrial events were regulated by the relative positions of the planets to each other and to one's local horizon, that is, by those positions at particular moments in time. In a cosmos of moving planets, astrological meaning arose only in time; and time, itself generated by the movement of the Sun, was public. To know time, even to set a mechanical clock, required the astrologer to leave the private, interior space and engage with time outside in the public space.

---

* I thank participants at the *XXXIV Symposium of the Scientific Instrument Commission* in Turin for their questions and comments, and especially Jim Bennett, John Davis, Petr Hadrava, Efthymios Nicolaidis, Günther Oestmann, Gertrud Oswald, Lutz Pannier, Pedro Raposo and Sara Schechner for their help and comments on earlier drafts of this essay.

1 For my recent study of a series of sixteenth-century paper instruments designed to serve the astrologer working in private spaces, see Richard L. Kremer, "Playing with Geometrical Tools: Johannes Stabius's *Astrolabium imperatorum* (1515) and its Successors", *Centaurus* 58 (2016), pp. 104–134.

 | DOI:10.1163/9789004381438_010

FIGURE 8.1 The astrologer in private space with his books and armillary sphere; from *Kalender*, Augsburg, 1539, title page.
COURTESY BAYERISCHE STAATSBIBLIOTHEK, MUNICH, RES/4 EPH.ASTR. 99 T.

Cultural historians have long studied societies' time consciousness, especially the shifts from private time to publicly controlled time as demanded by religious practices, merchants, markets and factories, military establishments, schools and railroads.[2] Each case required rules for standardizing time and

2 For an introduction to this massive literature, see Gerhard Dohrn-van Rossum, *History of the Hour*, trans. by Thomas Dunlap, Chicago, University of Chicago Press, 1996.

technologies and experts for keeping and communicating time. In this chapter, I want to consider another context for standardizing time, viz., the astrological practice of associating a particular planet with each of the 168 unequal hours in the week. Called "planetary hours", this convention of timekeeping went back to Greek antiquity. Ptolemy had referred to the planets as "chronocrators" or "masters of time", arguing that each planet ruled parts of the zodiac as well as its assigned hours.[3] The scheme divided daytime, from sunrise to sunset, into twelve hours of equal duration and night-time into twelve hours of equal duration. Only at the equinoxes are night and day hours of equal length. In the summer, day hours are longer than night hours; the reverse occurs during the winter. Mechanical clocks, sundials and tables of seven rows by 24 columns would be devised to order this astrological time, which was too complex to remember. Astrological time required material aids for its practice.

In the sixteenth century, an era marked by a profusion of mechanical clocks providing uniform, astronomical time, with each hour of each day having exactly the same duration, astrological time also found its place in public timekeeping technologies. This chapter will examine selected examples of astrological timekeeping in medieval and early modern public spaces; it cannot provide a comprehensive survey. We will begin with a unique sundial in Görlitz made around 1550, will then consider other examples of public astrological timekeeping provided by mechanical clocks or sundials, and will conclude by considering how some sixteenth-century dialists explored the mathematics of unequal hours on vertical sundials.

The Görlitz Arachne

If today you enter Görlitz, a small city in southeastern Germany on the Neisse River with much of its handsome early modern architecture intact, and walk to the Untermarkt, you will encounter the former Ratsapotheke, a large five-storey house of Renaissance design (*Fig. 8.2*). Looking up, you will see on its southern façade a recently restored fresco depicting two large sundials each about 3.2 × 1.2 m, entitled "Solarium" on the left and "Arachne" on the right. At the far left, a Latin verse describes the latter:

3 Claudius Ptolemy, *Tetrabiblos*, ed. and trans. by Frank Egleston Robbins, "Loeb Classical Library 435", Cambridge MA, Harvard University Press, 1940 (reprint: London, William Heinemann Ltd., 1980), pp. 452–457; Auguste Bouché-Leclercq, *L'astrologie grecque*, Paris, Leroux, 1899, pp. 491–506.

FIGURE 8.2 The Solarium (left) and Arachne (right) on the Görlitz Ratsapotheke, photographed 2010.
COURTESY CREATIVE COMMONS.

> The planetary hours are displayed by green lines
> The houses by red and white areas.
> With black threads the Sun's altitude is woven
> With this order, red, vertical lines divide its azimuth.[4]

Below the Arachne, a list of dates records the many restorations of these dials, from 1550 ("Scultetus invenit") through 1558, 1719, 1830, 1843, 1853, 1867, 1876, 1889, 1926, 1936, 1973 and finally 2000. Unlike dials carved in stone, delicate lines painted in plaster have poorly survived Görlitz's weather. We need not assume that the earlier dials looked like today's versions. However, the earliest surviving sketch of the dials, made after the 1830 restoration, includes all the features seen today; and presumably the Arachne's Latin verse is unchanged since 1550.[5] We may not be able to draw geometrical conclusions about the

4 "Hora planetarum viridi est expressa colore / Et rubra oscendunt albaque plana domus. / Stamine deinde nigro contexta ascensio solis / Quo rutila asimuth ordine norma secat."

5 Walther Zimmermann, "Die Sonnenuhren an der Ratsapotheke in Görlitz", *Neues Lausitzisches Magazin* 111 (1935), pp. 142–168, Abb. 3–7.

1550 version by measuring today's lines, but we can, I think, discuss the original functions of the Görlitz dials based on the surviving evidence.[6]

Although often attributed to Bartholomaeus Scultetus (1540–1614), Bürgermeister of Görlitz, friend of Tycho Brahe and well-known astronomer who in 1572 would publish a sophisticated treatise on dialing,[7] the Görlitz dials of 1550 apparently predate Bartholomaeus. In 1560, a local source attributed one dial to Bartholomaeus's older brother, Zacharias (1530–1560). However, recent research has found that the leftmost house, bearing the Solarium, was connected only after 1565. We might guess, therefore, that Zacharias designed the Arachne in 1550, Bartholomaeus the Solarium sometime after 1565.[8]

The Solarium marks equal hours (24 per day, always of the same length), the place of the Sun in the ecliptic over the course of the year, day lengths, Italian hours (24 equal hours per day counted from sunset) and Babylonian hours

6 According to a report on the restoration in 2000, the entire façade of the building, including its angle of deviation from the east-west direction, has changed since the sixteenth century. No lines of the original dials remain; the latest restorers recomputed the lines, struggling with the results of earlier reconstructions. "An analysis of the net of lines [of the Arachne] revealed chaos. The azimuth and altitude lines contradicted each other and did not match the existing gnomon; from those lines no theoretically correct gnomon could be derived. This raised the question of whether such serious errors were made during the renovation of 1926/36 or simply tolerated." Lutz Pannier, "Historische Sonnenuhren in Görlitz, Teil 2", *Görlitzer Magazin* 16 (2003), pp. 66–77: 70.

7 Bartholomaeus Scultetus, *Gnomonice*, Görlitz, Ambrosius Fritsch, 1572. This work does not mention the Görlitz dials on the Untermarkt.

8 A Görlitz chronicle, written in 1560 and first published in 1719, describes the death of Zacharias Scultetus in 1560: "he knew many languages, was excellent in Greek, Latin, German, Slavic. His death was a great loss but perhaps his brother Bartholomew can make up for it. Zacharius made two sundials, one in the market on the house of Hoffmann, the other on the building of his son-in-law, Michael Scultetus, both drawn with great skill." Martin Mylius, "Annales Gorlicenses ab anno P.C.N. MCXXXI continuati usque ad anno MDCLI", in Christian Hoffmann, Christoph Manlius (eds.), *Scriptores rerum Lusaticarum antiqui et recentiores*, Leipzig, Richter, 1719, i/2: pp. 35–36. For previous studies of the Görlitz dials, see Richard Jecht, "In welchen Häusern wohnten die berühmten Görlitzer?", *Neues Lausitzisches Magazin* 219 (1913), pp. 214–222: 219–220; Otto Eichler, "Die Sonnenuhren an der G. Struves Apotheke", *Neuer Görlitzer Anzeiger* 291 (1926), p. 3, and suppl., p. 1; Hellmut Winkler, "Astronomische Uhren in Görlitz", *Mitteilungen der Einstein-Sternwarte Görlitz* 3 (1956), pp. 1–19; Arnold Zenkert, *Katalog der ortsfesten Sonnenuhren in der DDR*, Berlin-Treptow, s. publ., 1984, p. 7; Christine Reidel, "Die Sonnenuhr an der Görlitzer Ratsapotheke", *Sterne und Weltraum* 35 (1995), pp. 64–65; Arnold Zenkert, "Die Arachne von Görlitz: Dokument der Astrologie", in Jürgen G.H. Hoppmann (ed.), *Melanchthons Astrologie*, Wittenberg, Drei Kastanien Verlag, 1997, pp. 21–23.

FIGURE 8.3 The Görlitz Arachne, photographed 2010.
COURTESY CREATIVE COMMONS.

(24 equal hours per day counted from sunrise).[9] Such timekeeping, based on the uniform astronomical hour, served civic activities and was provided by many Renaissance sundials and mechanical clocks, both public and private.

The Arachne, with functions described by the Latin verse, marks what we might call astrological time (*Fig. 8.3*). The alternating red and white swatches

9 For various forms of Renaissance hour-reckoning, see Dohrn-van Rossum, op. cit. (n. 2), pp. 113–117. An earlier Latin inscription for the Solarium, no longer preserved on the Ratsapotheke, referred to Nuremberg hours, yet another way of counting hours starting at sunrise, with differing numbers of hours of equal length allotted to day and night, over the course of the year. Since none of the earlier known lines on the Solarium could have marked Nuremberg hours, the 2000 restorers speculated that the Scultetus brothers may have thought that Nuremberg hours are identical to Babylonian hours, which are now and were earlier depicted on the Solarium. See Zimmermann, op. cit. (n. 5), pp. 153–154; Pannier, op. cit. (n. 6), p. 69.

mark the entry of the Sun's shadow into the six daytime houses of the horoscope. The Latin terms Dom[us] are inscribed on the wall to denote these houses. The Sun rises into Domus 12, moves next into Domus 11, then into Domus 10. At noon it enters Domus 9. As the Sun sets below the horizon, it exits Domus 7. Each house had its unique astrological significance, meanings defined already in Greek antiquity.[10] A glance at the Arachne would allow astrologically attuned observers to know, at that instant, where the Sun was in the horoscope. For example, the Sun rules husbands; a wife might encourage her husband to embark on a journey when the Sun enters House 9, the house of travel.

The Arachne also indicates the planetary hours, which require what at first glance appears to be a confusing net of lines and marks on the dial, hence its name. Grey lines mark the unequal hours that divide each day, from sunrise to sunset, into twelve equal segments of time. Greek astrologers had assigned planets to rule each unequal hour over the course of a week, thereby creating the planetary hours.[11] The algorithm generating these hourly rulers is simple: list the planets from the slowest to the fastest (Saturn, Jupiter, Mars, Sun, Venus, Mercury, Moon) and associate each successive planet with each successive hour, starting with Saturn for the first hour of the day of Saturday and continuing to the next successive day. But counting out 168 planetary hours over a week is cumbersome; many sixteenth-century imprints reduced the scheme to tables (*Fig. 8.4*).

The Görlitz Arachne ostensibly makes things easier by labelling the planetary hours with twelve columns, one for each of the twelve daytime hours, of the seven planetary symbols. The symbol for Saturn (♄) appears at the top left corner of the Arachne. When the gnomon's shadow falls in the leftmost region marked by the grey lines (the first hour of the day) on Saturday, Saturn rules the hour. On Sunday, that hour is ruled by the next symbol to the right, Venus (♀); on Monday, by the next symbol to the right, Mars (♂), etc. If the Sun is in the second hour, the second column of symbols denotes the hours, Jupiter (♃) for Saturday, Mercury (☿) for Sunday, Mars for Monday, etc. The photograph in *Fig. 8.2* shows the solar shadow in the sixth hour of the day, ruled by Mercury on Saturday, by Mars on Sunday, etc. On sunny days, a simple glance at the Arachne enables the viewer to know the planetary hour at any instant.

The Arachne provides two further indications of the Sun's position over the course of the day. A set of thick, red vertical lines mark in 5° intervals the

10 The classic house interpretations, starting with House 1, include the subject's future life, business, brothers, parents, children, illness, marriage, death, travel, honors, friends and enemies. See Bouché-Leclercq, op. cit. (n. 3), pp. 280–288.

11 See Oestmann's chapter in this volume, *infra*, pp. 86–103.

Dies artificialis.	Horæ inæquales diurnæ, & nocturnæ.												Nox artificialis.
	1	2	3	4	5	6	7	8	9	10	11	12	
Dominicus.	☉	♀	☿	☾	♄	♃	♂	☉	♀	☿	☾	♄	Mercurij.
Lunæ.	☾	♄	♃	♂	☉	♀	☿	☾	♄	♃	♂	☉	Iouis.
Martis.	♂	☉	♀	☿	☾	♄	♃	♂	☉	♀	☿	☾	Veneris.
Mercurij.	☿	☾	♄	♃	♂	☉	♀	☿	☾	♄	♃	♂	Sabbathi.
Iouis.	♃	♂	☉	♀	☿	☾	♄	♃	♂	☉	♀	☿	Dominicæ.
Veneris.	♀	☿	☾	♄	♃	♂	☉	♀	☿	☾	♄	♃	Lunæ.
Sabbathi.	♄	♃	♂	☉	♀	☿	☾	♄	♃	♂	☉	♀	Martis.

FIGURE 8.4 Planetary hours in tabular format; from Oronce Finé, *De duodecim caeli domiciliis et horis inequalibus*, Paris, 1553, fol. 21r.

azimuth of the Sun, i.e. its angle along the horizon either east or west of the north-south (or meridian) line from 0° to 90°. And a set of black hyperbolic lines, open downward, mark the altitude of the Sun, also at 5° intervals from 0° to 60°. As far as I can tell, these solar positions, relative to the Görlitz horizon, have neither astrological nor time-telling significance. On the other hand, knowing the azimuth and altitude of the Sun would enable anyone to set the rete of a hand-held astrolabe to the position of the celestial sphere for the time in question. And the presence of these lines on the dial, crossing the astrologically relevant lines, contributes dramatically to the image of a spider's web on the façade of the Ratsapotheke.

Indeed, the term arachne, like the complexity of the dials' lines, might well have been intended to celebrate the humanistic pretentions of the Scultetus brothers. In *De architectura* (IX, 8) Vitruvius had listed several types of sundials, including the "arachne", whose invention he attributed to Eudoxus or Apollonius. Derived from the Greek αραχνη, arachne referred to a spider's web.[12] Vitruvius's text provides neither descriptions nor diagrams and scholars have long debated what this Greek arachne dial may have been. The *editio princeps* of Vitruvius appeared in 1486;[13] the first edition in Italian (1521), translated

12 The Greek loan word "arachne" is very rare in Latin, with only one other attested usage, in the *Liber medicinalis* of Serenus Sammonicus (d. 212), line 957. See *Thesaurus Linguae Latinae*, s.v. "arachne."

13 Marcus Vitruvius Pollio, *De Architectura*, Rome, [Printer of Vitruvius], 1486, fol. 77v.

with a commentary by the Milanese military engineer and painter Cesare Cesariano, struggled to explain the sundial web:

> The renowned astronomer [astrologo] Eudoxos [...] discovered [trovò] the clock spider [Aranea Horologica]. We can say that this had the form of a Sun that irradiates rays from the centre to the circumference—as many [rays] as the number of the days in the year. It looks like a spider from the centre of whose body a large number of feet departs in a circular manner or, even better, [it looks like] a spider's web composed of circles and divided into small squares. The gnomon was located in it in the proper manner.[14]

The first German edition (1548), with shorter commentary, also avoided geometry; the Sun clock was called arachne "because of the manifold crossing [Durchschrenckung] of the lines".[15] Not until Daniele Barbaro's lengthy commentary (1556) would an author speculate on the function of the lines. "The arachne could be a clock whose hour lines are crossed by circles showing the solar altitude according to shadow lengths or the height of the gnomon placed in the hemisphere of the circle", probably a reference to a spherical dial with inscribed hour lines and circular day curves, a common design among the surviving Greek and Roman dials (*Fig. 8.5*).[16] Subsequent sundial historians have speculated that Eudoxus's arachne could have been a planar, vertical dial or a mobile metal net turning within a hemispheric dial not unlike the rete on an astrolabe.[17] In any case, it seems quite likely that Zacharias Scultetus borrowed the name for his Görlitz dial from Vitruvius.

---

14 Vitruvius, *De Architectura libri dece[m], traducti de latino in vulgare*, Como, Gotardus da Ponte, 1521, fol. 158v.

15 Vitruvius, *Vitruvius teutsch*, trans. by Walther Ryff, Nuremberg, Johann Petreius, 1548, fol. 283r.

16 Vitruvius, *Dell'architettura*, trans. by Daniel Barbaro, Venice, Francesco Marcolini, 1556, p. 236. For the spherical dial of *Fig. 8.5*, see Sharon L. Gibbs, *Greek and Roman Sundials*, New Haven, Yale University Press, 1976, p. 133. For photographs of similar examples, see Karlheinz Schaldach, *Die antiken Sonnenuhren Griechenlands: Festland und Peloponnes*, Frankfurt am Main, Deutsch, 2006, pp. 91, 95, 96–97, 100, 104, 107, 110, 122, 125, 127, 132–136, 138 and 142.

17 Joseph Drecker, *Die Theorie der Sonnenuhren*, Berlin, W. de Gruyter, 1925, p. 55; Jean Soubiran (ed.), *Vitruve, De l'Architecture*, Paris, Les Belles Lettres, 1969, pp. 248–251; Gibbs, op. cit. (n. 16), pp. 59–61; Jérôme Bonnin, *La Mesure du temps dans l'antiquité*, Paris, Les Belles Lettres, 2015, pp. 110–119. For a recent proposal that a dial from fourth-century BCE Amphiareion provides the prototype for Eudoxus's "invention", see Karlheinz Schaldach, "The Arachne of the Amphiareion and the Origin of Gnomonics in Greece", *Journal for the History of Astronomy* 35 (2004), pp. 435–445; Schaldach, op. cit. (n. 16), pp. 31–32 and

FIGURE 8.5 A model for Vitruvius's "arachne"? Greek spherical dial showing eleven hour lines and three day curves, height 27.5 cm, found at Lanuvium in 1891 (British Museum 1893,0713.13).

Planetary Hours in Public Space

Displaying unequal, as well as equal, hours on public timekeeping devices, as occurs on the Görlitz Ratsapotheke, was not uncommon in early modern Europe. Emmanuel Poulle has suggested that this dual function on mechanical clocks, both large and small, mirrored the functions of the astrolabe, a small, hand-held instrument that provides users with both equal and unequal hours.[18] The earliest extant text on the astrolabe, by the sixth-century

116–121. For an exceedingly speculative reconstruction of Eudoxus's arachne not as a sundial but as an astronomical observing instrument and a computational tool capable of extracting square and cube roots, see Erkka Maula, Eero Kasanen, Jorma Mattila, "The Spider in the Sphere: Eudoxus' Arachne", *Philosophia, Yearbook of the Research Center for Greek Philosophy at the Academy of Athens* 5–6 (1975–1976), pp. 225–258.

18 Emmanuel Poulle, "L'horlogerie a-t-elle tué les heures inégales?", *Bibliothèque de l'École des Chartes* 157 (1999), pp. 137–156.

Alexandrian philosopher Ioannes Philoponus, describes the geometry and use of the "hour lines". Drawn only below the horizon to avoid, as Philoponus explained, confusion with the circles marking azimuth and altitude above the horizon, these looping curves divide the nocturnal arc of the Sun, as well as its diurnal arc, into twelve equal parts and thus mark the unequal hours.[19] Stöffler's widely circulated *Elucidatio fabricae ususque astrolabii* (1513) neatly instructed readers how to draw the "temporary or unequal" hour lines on the tympanum (*Fig. 8.6*). Equal hours are marked by a rotation of the rete through one twenty-fourth of the circle, or 15° around the rim of the astrolabe. If one knows the longitude of the Sun for the day in question and sets the rete to the position of the celestial sphere at any given time, the astrolabe enables users to read that time in both equal and unequal hours.[20] The classical, planispheric astrolabe, we might say, tells time in two registers.

This double time-telling also emerged very early in the large mechanical clocks that started appearing on civic buildings in fourteenth-century Europe (*Table 8.1*).[21] In most of these machines, mechanical works rotated an eccentric zodiac (an abbreviated rete) one turn per day, with attached pointers indicating the solar and lunar places in the zodiac. The fixed tympanum below the zodiac marks the horizon and curves for the unequal hours.

---

19 Alfred Stückelberger, *Ioannes Philoponus, De usu astrolabii eiusque constructione*, Berlin, W. de Gruyter, 2015, pp. 11, 23–24; Claude Jarry, *Jean Philopon, Traité de l'astrolabe*, Paris, Les Belles Lettres, 2015, pp. 9, 17–19. The earliest extant Islamic and European astrolabes generally include lines for the unequal hours. See David A. King, "The Earliest Known European Astrolabe in the Light of other Early Astrolabes", *Physis* 32 (1995), pp. 359–404. See also Schmidl's chapter in this volume, *infra*, pp. 5–10.

20 Johann Stöffler, *Elucidatio fabricae ususque astrolabii*, Oppenheim, Jacob Köbel, 1513, fol. 7r-v. For a useful introduction to the astrolabe, see James Evans, *The History and Practice of Ancient Astronomy*, New York, Oxford University Press, 1998, pp. 141–158.

21 The list in *Table 8.1*, representative but not exhaustive, has been drawn from E. Chateau, *L'horloge astronomique de Saint-Jean à Lyon*, Paris, L. & A. Cresson Frères, 1901; Alfred Ungerer, *Les horloges astronomiques et monumentales*, Strasbourg, Chez l'auteur, 1931; Klaus Maurice, *Die deutsche Rädeuhr*, Munich, C.H. Beck, 1976, 2 vols.; Günther Oestmann, *Die astronomische Uhr des Straßburger Münsters*, Stuttgart, Verlag für Geschichte der Naturwissenschaften und der Technik, 1993; Charles-Henri Eyraud, "Horloges astronomiques au tournant du XVIIIe siècle", PhD dissertation, Université de Lyon, 2004; Manfred Schukowski, *Wunderuhren*, Schwerin, Thomas Helms, 2006; Emmanuel Poulle, "L'horloge astronomique de Bourges", *Bulletin de l'Association nationale des collectionneurs et amateurs d'horlogerie ancienne* 113 (2008), pp. 21–32; Christine Magin, *Die Inschriften der Stadt Stralsund*, Wiesbaden, Dr. Ludwig Reichert Verlag, 2016, pp. 74–79. Several of these astronomical clocks show the astrolabe plate as projected from the north pole, which places the daytime sky at the top of the dial and eases reading of the daytime planetary hours. See Zdeněk Horský, "Astronomy and the Art of Clockmaking in the Fourteenth, Fifteenth and Sixteenth Centuries", *Vistas in Astronomy* 9 (1967), pp. 25–34.

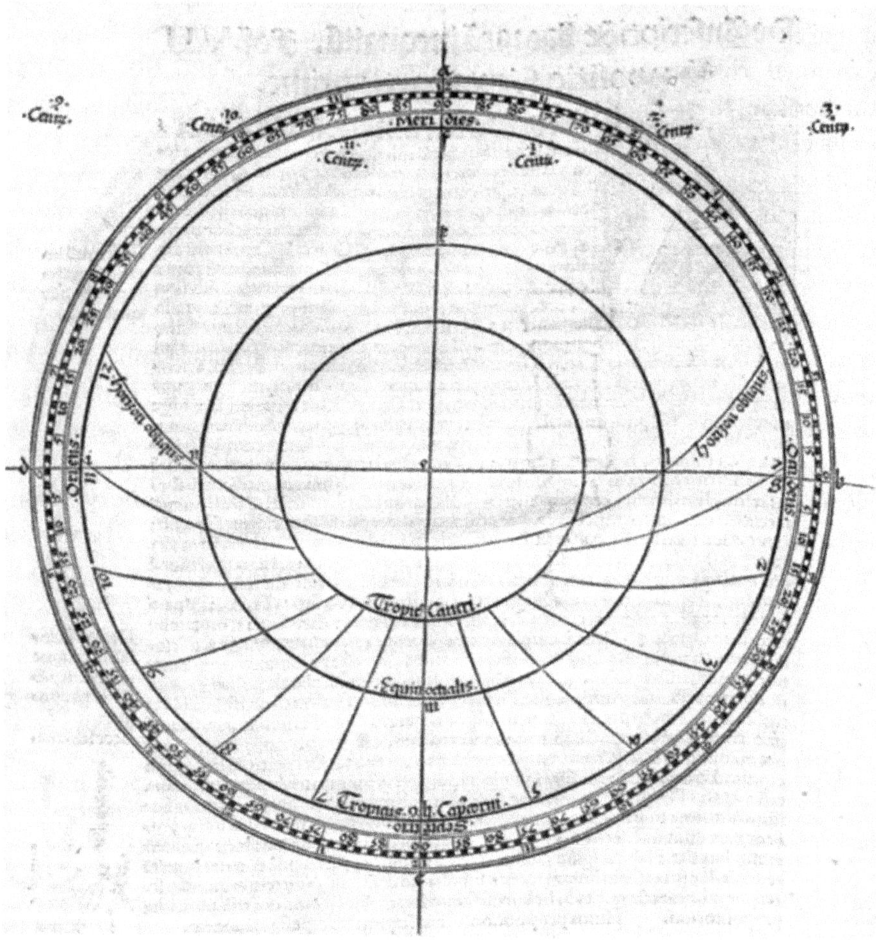

FIGURE 8.6 Astrolabe tympanum with curves marking unequal hours, numbered 1 to 12; from Johann Stöffler, *Elucidatio fabricae ususque astrolabii*, Oppenheim, 1513, fol. 7v. COURTESY BAYERISCHE STAATSBIBLIOTHEK, MUNICH, RES/2 MATH.A. 94.

Although the astrolabe plates on these clocks mark lines for the unequal hours, the widths of the metal circles representing the ecliptic and the markers for the Sun often obscure much of the underlying tympani. In most cases, it would have been difficult for a viewer, standing on the ground, to read the unequal hours of the day. Indeed, we might guess that the astrolabes on these public clocks served more of a representational function, illustrating the daily

TABLE 8.1 Public astronomical clocks marking unequal hours, before 1600. Dates refer to the first construction of a clock that might subsequently have been modified or replaced, numbers in brackets in the second column to successive clocks in given buildings

Date	Location	Maker	Technique
1353	Strasbourg cathedral (1)	Unknown	Astrolabe
1383	Lyon cathedral	Unknown	Astrolabe
1390	Doberan cathedral	Nicolaus Lillienfeld (?)	Astrolabe
1394	Stralsund cathedral	Nicolaus Lillienfeld	Astrolabe
1405	Bern tower	Unknown	Astrolabe
1410	Prague cathedral	Unknown	Astrolabe
1424	Bourges cathedral	Jean Fusoris	Astrolabe
1425	Lund cathedral	Nicolaus Lilienfeld (?)	Astrolabe
1470	Danzig cathedral	Hans Düringer	Astrolabe
1472	Rostock cathedral	Hans Düringer (?)	Planetary hours disk
1473	Mantua tower	Bartolomeo de Manfredi	Astrolabe
1540	Münster cathedral (2)	Johannes von Aachen	Astrolabe + slots
1555	St Omer cathedral	Pierre Engueran	Astrolabe?
1561	Lübeck cathedral (2)	Matthias von Oß	Astrolabe + slots
1571	Strasbourg cathedral (2)	Conrad Dasypodius	Astrolabe
1581	Ulm town hall (2)	Isaak Habrecht	Astrolabe
1594	Augsburg cathedral	Tobias Klieber	Astrolabe

rotation of the cosmos, than a practical time-telling function, at least for the unequal hours.[22]

Occasionally public astronomical clocks were designed for what, to our eyes, might seem as easier reading of unequal hours. For example, the late Augsburg clock presented the unequal hours, labelled "planetary hours", by a separate dial with the daily rotation of the Sun marked by a pointer rather than a full rete. Beneath the dial was a table identifying the planetary rulers (*Fig. 8.7*). If viewers knew the solar longitude for the day in question, thereby determining the radius of the solar pointer, they could identify the unequal hour and then find its ruler by referring to the table below the dial.

22 Many sixteenth- and seventeenth-century fancy German table clocks, displayed in private spaces, include astrolabe dials that undoubtedly served both representational and practical functions. See Maurice, op. cit. (n. 21).

FIGURE 8.7 Tobias Klieber's drawing, *c.* 1600, of the astrolabe dial and table for finding planetary hours on the Augsburg public clock.
COURTESY BAYERISCHE STAATSBIBLIOTHEK, MUNICH, CGM 4189, FOL. 7R.

Other two-step processes for reading unequal hours can be found on the astronomical clocks in Lübeck and Münster. On either side of the central dial presenting the astrolabe is a column of slits to indicate the twelve day (left) and night (right) unequal hours. Behind each slit is a mechanically driven wheel

FIGURE 8.8 Mechanical dial for planetary hours on the Rostock astronomical clock, *c.* 1472. COURTESY MANFRED SCHUKOWSKI, *WUNDER UHREN*, SCHWERIN, 2006, P. 96.

containing on its edge the planetary names. Arranged in order of the planetary hours and advanced one planet each day (all twelve wheels ride on a common shaft), these slits announce the given planets ruling each of the unequal hours over the course of the day.[23]

A one-step mechanical solution for the planetary hours had appeared a century earlier on the Rostock clock (*Fig. 8.8*). Clockmaker Hans Düringer fixed at the end of the hour pointer on the main dial a small dial divided into 28 sectors, showing the signs and names of the seven planets, repeated four times in counterclockwise sequence of the planetary hours. A red pointer rotates

23 For illustrations and a description, see Schukowski, op. cit. (n. 21), pp. 70, 72, 86.

counterclockwise on this dial, indicating the planetary sign at its one end and the planetary name at the other. In one day, the hour pointer rotates once around the large dial marked in 24 hours and the red pointer completes $^{24}/_{28}$ of a revolution with respect to the hour pointer. The red pointer thus indicates the 24 planetary hours of the day. As Manfred Schukowski has shown, Düringer achieved this movement very simply by attaching a 24–toothed wheel on the reverse side of the small dial, weighted so that it does not rotate with respect to the ground as the hour pointer rotates. A 28–toothed wheel, placed at the centre of the small disk, engages the other wheel and drives the red pointer $^{24}/_{28}$ of a revolution per day.[24] Previous scholars, however, have not noted that the Rostock mechanism yields "planetary hours" of constant length, identical to the civil hours. It does not give astrological time as it would have been understood in sixteenth-century Europe, which might explain why Düringer's mechanism apparently was not replicated on any later astronomical clock.[25]

Although sundials appeared on early modern European public buildings far more frequently than did the elaborate astronomical clocks, they rarely offered planetary hours. I have found only two such public dials made prior to 1600, the Arachne in Görlitz and a dial on the south transept wall of the Katharinenkirche in Oppenheim, a free imperial city from the thirteenth century and from the fourteenth an administrative centre on the Rhine for the Palatinate.[26] One of the largest Gothic churches along the Rhine between Cologne and Strasbourg, the Katharinenkirche was constructed from the thirteenth through the fifteenth centuries. In 1586, two sundials were inscribed (rather than painted) into the red sandstone on either side of a window.[27]

24 M. Schukowski, "Astronomische Monumentaluhren in Kirchen", in Jürgen G.H. Hoppmann (ed.), *Melanchthons Astrologie*, Wittenberg, Drei-Kastanien-Verlag, 1997, pp. 24–26; Schukowski, op. cit. (n. 21), pp. 96–99; Helge Tideke, "Die geniale Scheibe der Planetenstunden an der astronomischen Uhr in St. Marien zu Rostock", *Mitteilungen der deutschen Gesellschaft für Chronometrie* 143 (2015), pp. 27–28.

25 Had a sundial or an astrolabe dial been attached to the Rostock clock, viewers could have read the unequal hour from these devices and then used the mechanical dial to determine the planetary ruler for that hour. But as far as I know, the Rostock clock did not offer unequal hours.

26 Cf. Denis Savoie, *Recherches sur les cadrans solaires*, Brepols, Turnhout, 2014, p. 128.

27 Ernst Zinner, *Alte Sonnenuhren an europäischen Gebäuden*, Wiesbaden, F. Steiner, 1964, Abb. 38; Siegrid Düll, *Die Inschriften der Stadt Oppenheim*, Wiesbaden, Ludwig-Reichert-Verlag, 1984, pp. 90–91; Norbert Becher, "Die Sonnenuhren an der Katharinenkirche", *Oppenheimer Hefte* 16 (1998), pp. 2–49. In 1689, a fire badly damaged the Katharinenkirche; it was not rebuilt until the nineteenth century. Both dials were renewed in 1935. It is impossible to know the original disposition of the Oppenheim dials.

FIGURE 8.9 The Oppenheim dials for planetary (left) and civil hours (right), 1586, restored 1935, photographed 2008.
COURTESY CREATIVE COMMONS.

As in Görlitz, the Oppenheim dial, denoted as a "Horologium planetarum," is paired with a dial for the civil hours (*Fig. 8.9*). Thousands of sundials can be found on European public buildings constructed before 1600; yet as far as I know only two indicate the planetary hours. Unfortunately no sources have been preserved that might reveal why the Scultetus brothers or the unknown dial-maker in Oppenheim created special sundials for astrological time.

The Mathematics of Astrological Time on Vertical Sundials

Although the Oppenheim and Görlitz dials share a similar design, the former is simpler (*Fig. 8.10*), providing only the planetary hours, labelled in seven rows with signs for the planets (names are given for the leftmost entry of each row). Due to the wall angled southward at the left that blocks the evening Sun, the Oppenheim dial marks only hours 1 through 10; the final two hours of each day are not represented at the right side of the dial. Hyperbolae mark the day curves of the tip of the shadow for the three zodiacal signs north and south of the equinox. The current Oppenheim gnomon is too long, since the tip of its shadow far exceeds the day line for the summer solstice. The Oppenheim dial does not provide lines for the astrological houses. And the Oppenheim dial is declined by about −10°, i.e. its wall faces slightly eastward (the equatorial day

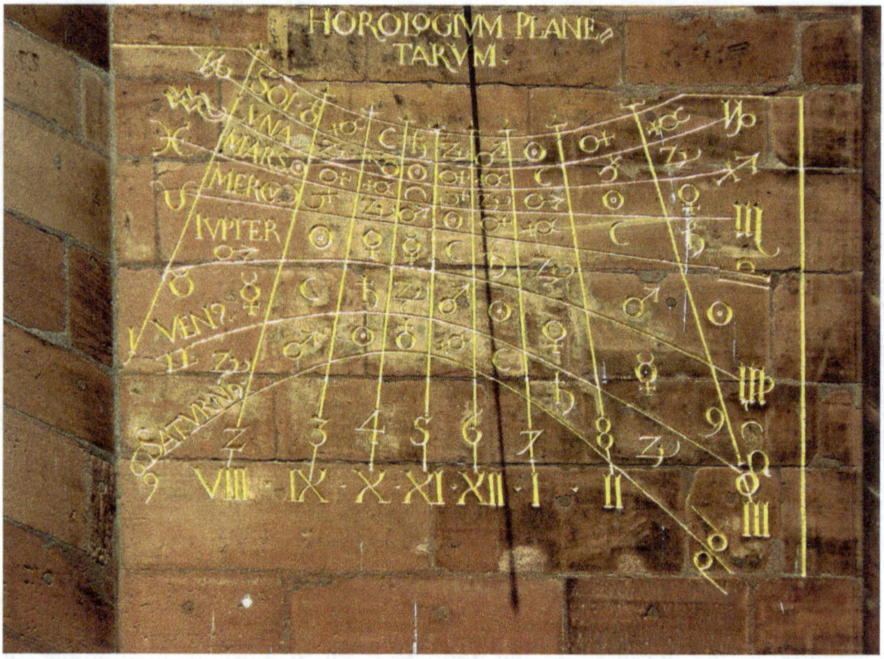

FIGURE 8.10 The Oppenheim "Horologium planetarum", photographed 2008. COURTESY CREATIVE COMMONS.

line is not horizontal but tilts downward toward the east). The Arachne, as currently painted, is not declined.[28]

Like mechanical clocks with astrolabe dials, the Görlitz and Oppenheim sundials mark unequal hours of what sundial historian Denis Savoie has called the "first type". They divide the Sun's diurnal arc into twelve equal segments. For a vertical dial with a perpendicular gnomon, Savoie has shown that these lines of unequal hours can be computed in rectangular coordinates:

$$x = a\frac{\cos D \sin T - \sin\phi \sin D \cos T + \cos\phi \sin D \tan\delta}{\sin D \sin T + \sin\phi \cos D \cos T - \cos\phi \cos D \tan\delta}$$

28 The earliest nineteenth-century drawing of the Görlitz Arachne was made from the street, looking up at the dial. Although the parallel azimuth lines remain parallel in the drawing (a photograph from that perspective would not preserve the parallel pattern), the unequal hour lines have been spread to greater angles from the meridian line than the actual angles on the wall, a feature a photograph would create. Without knowing the vantage point of the drawing, it becomes impossible to know the actual angles of the dial's original design.

$$y = a\frac{-\cos\phi\cos T - \sin\phi\tan\delta}{\sin D\sin T + \sin\phi\cos D\cos T - \cos\phi\cos D\tan\delta}$$

where a is the length of the gnomon, located at the origin, D the declination of the dial, δ the solar declination, ϕ the geographical latitude, and:

$$T = k\frac{\arccos(-\tan\phi\tan\delta)}{6}$$

where k denotes the unequal hour, with $k = 0$ corresponding to the sixth hour, $k = 1$ the seventh hour, $k = -1$ the fifth, etc.[29] Sixteenth-century dialists, of course, would have worked with analemma methods rather than sines and cosines.[30] I do not find unequal hours treated in the widely known gnomonic works of Sebastian Münster or Oronce Finé. Instructions for constructing unequal lines on a vertical dial do appear in Christoph Clavius's massive *Gnomonices libri octo* (1581).[31]

Interestingly, neither the Oppenheim nor the Görlitz dial provides unequal hours of Savoie's "second type", i.e. durations defined by the rising of 15° of the ecliptic rather than one-twelfth of the Sun's diurnal arc. I shall thus call the "second type" unequal hours "ecliptic". They are unequal over the course of a single day as well as over the course of the year (*Table 8.2*). Sixteenth-century dialists did not consistently treat these two types of unequal hours, disagreeing conceptually and terminologically. Lacking space here to comprehensively survey these ambiguities, let me mention only several examples.

As Joseph Drecker noted, Johannes de Sacrobosco (*c.* 1195–*c.* 1256) seems to have been among the earliest writers to define ecliptic unequal hours: "However short or long the day or night may be, six signs rise by day and six by night [...] since a natural hour is the space of time in which half a sign [15° of the ecliptic] rises, there are twelve natural hours in each artificial day

29 Savoie, op. cit. (n. 26), p. 133; Denis Savoie, *La gnomonique*, 2nd enlarged ed., Paris, Les Belles Lettres, 2007, pp. 297–300.

30 I cannot here discuss analemma methods for solving the spherical trigonometry of dialing. Cf. Gustav Bilfinger, *Die Zeitmesser der antiken Völker*, Stuttgart, Carl Liebich, 1886; Rocco Sinisgalli, Salvatore Vastola, *L'analemma di Tolomeo*, Florence, Cadmo, 1992; Armin Zenner, "Über das Analemma", *Jahresschrift der deutschen Gesellschaft für Chronometrie* 39 (2000), pp. 211–226; Nathan Sidoli, "Heron's *Dioptra* 35 and Analemma Methods: An Astronomical Determination of the Distance between Two Cities", *Centaurus* 47 (2005), pp. 236–258.

31 Sebastian Münster, *Compositio horologiorum*, Basel, Petrus, 1531; Oronce Finé, *De solaribus horologiis et quadrantibus*, Paris, Gulielmum Cauellat, 1560; Christoph Clavius, *Gnomonices libri octo*, Rome, Franciscus Zanettus, 1581, pp. 236–237.

TABLE 8.2 Durations of ecliptic unequal hours in terms of civil hours, as a function of solar longitude for a geographical latitude of 51° (Görlitz), obliquity of 23° 30′ and durations of "first type" unequal hours given in top row

Unequal hour	λ⊙=0°, 1:00h	λ⊙=30°, 1:10h	λ⊙=60°, 1:18h	λ⊙=90°, 1:22h
1	0:26	0:32	0:48	1:11
2	0:28	0:38	0:59	1:20
3	0:32	0:48	1:11	1:24
4	0:38	0:59	1:20	1:25
5	0:48	1:11	1:24	1:25
6	0:59	1:20	1:25	1:25
7	1:11	1:24	1:25	1:25
8	1:20	1:25	1:25	1:25
9	1:24	1:25	1:25	1:25
10	1:25	1:25	1:25	1:24
11	1:25	1:25	1:25	1:20
12	1:25	1:25	1:24	1:11

[Sun above the horizon], and so also in the night [Sun below the horizon]".[32] Sacrobosco did not employ the term "planetary" hours. His "natural" hours are of unequal durations since the rising times vary for any given fixed interval (e.g. half a sign) on the ecliptic. Tables of oblique ascensions present this relationship; I computed oblique ascensions to produce *Table 8.2*.

Astrolabe treatises expanded Sacrobosco's vocabulary and altered his meaning. Stöffler's *Elucidatio* defined the "natural" day as the return of the Sun to the meridian, a period divided into 24 "equinoctial" or "equal" hours. The "artificial" day, when the Sun was above the horizon for a given latitude, is divided into twelve "temporary, natural, unequal or planetary" hours, each of the same length. But since over the course of the year the duration of the "artificial day" varies, so too does the duration of the "unequal" hour. Stöffler explained: "the unequal hour is not called 'unequal' in relationship to the other hours of the same day, as they are all the twelfth part of it, but in relationship to the hours of another day".[33] Sacrobosco defined "second type" hours, Stöffler "first type".

32 "[…] cum hora [naturalis, a term not in all the manuscripts] sit spatium temporis quo medietas signi peroritur, in qualibet die artificiali, similiter in nocte, 12 sunt hore naturales". Lynn Thorndike, *The Sphere of Sacrobosco and Its Commentators*, Chicago, University of Chicago Press, 1949, pp. 104, 134. Cf. Drecker, op. cit. (n. 17), p. 73.

33 Stöffler, op. cit. (n. 20), fol. 35r.

The subsequent flood of sixteenth-century imprints on instruments, especially on astrolabes, generally described “first type” unequal hours. I have found only isolated examples of ecliptic unequal hours in these materials. An anonymous set of manuscript drawings of instruments and dials, linked to Nuremberg in the 1510s–1520s, includes a careful rendering of ecliptic unequal hour lines for an astrolabe (*Fig. 8.11*). Labelled “Figura domorum per regulam qua indicatas signum peroritur”, the drawing seems intended for marking the houses of a horoscope according to the “single longitude method” where house cusps are set by uniformly dividing the ecliptic into 30° segments starting at the ascendant. Unequal hours are not mentioned.[34] In his *Instrument Buch* of 1533, Petrus Apianus (Peter Bienewitz, 1495–1552) noted that daytime, from sunrise to sunset, is divided into twelve “planetary hours” and that one such hour is “not (as the common crowd thinks) as long as the other; rather the sixth hour (ending at noon) is in the longest days always longer than the first hour after sunrise” and the reverse holds in the shortest days. Apianus offered no geometrical explanation for these hours; but he presented lines on a quadrant that allow users to sight the solar altitude and read the ecliptic planetary hour directly with a string and pearl (*Fig. 8.12*).[35] Several decades later the Parisian mathematician Oronce Finé introduced a new paper instrument for the working astrologer (*Fig. 8.13*). Finé’s device was essentially an astrolabe rete and tympanum marked with large arcs for Regiomontanus’s house boundaries (see below) and swooping lines for the ecliptic unequal hours for lengthening days. Finé also tabulated, for the latitude of Paris, the lengths (in degrees of right ascension) of these unequal hours as a function of the ascendant.[36]

None of these texts, however, implements ecliptic unequal hours on a sundial. The earliest known published diagram for a horizontal dial with unequal hour lines of the second type appeared in 1644, in a book on dialing by the Spanish scholastic philosopher and mathematician, Juan de Caramuel y Lobkowitz (*Fig. 8.14*).[37] Given this paucity of examples of ecliptic unequal

34 Bayerische Staatsbibliothek, Münich, Cod.icon. 182, fol. 34r.

35 Petrus Apianus, *Instrument Buch*, Ingoldstadt, Peter und Georg Apian, 1533, sig. A3v; Kurt Scheuerer, “Zeitbestimmung bei Apian”, in Karl Röttel (ed.), *Peter Apian*, Eichstatt, Polygon, 1999, pp. 85–92: 91–92.

36 Oronce Finé, *De duodecim caeli domiciliis et horis*, Paris, Michael Vascosanus, 1553, fols. 22v–24r.

37 Juan de Caramuel y Lobkowitz, *Solis et artis adulteria*, Louvain, Andreas Bouvetius, 1644, p. 72. Only the double line in the centre, bulging to the left, correctly represents the sixth ecliptic unequal hour; the remaining bulging lines differ by one civil hour from the sixth hour line, a flaw discovered by Fer J. de Vries, “Nicola Severino Finds Important News About Ecliptical Planetary Hours”, *Compendium, Quarterly Journal of the North American Sundial Society* 16/1 (2009), pp. 15–26. For a mathematically correct representation of ecliptic unequal hour lines for a vertical sundial, see Drecker, op. cit. (n. 17), Abb. 104.

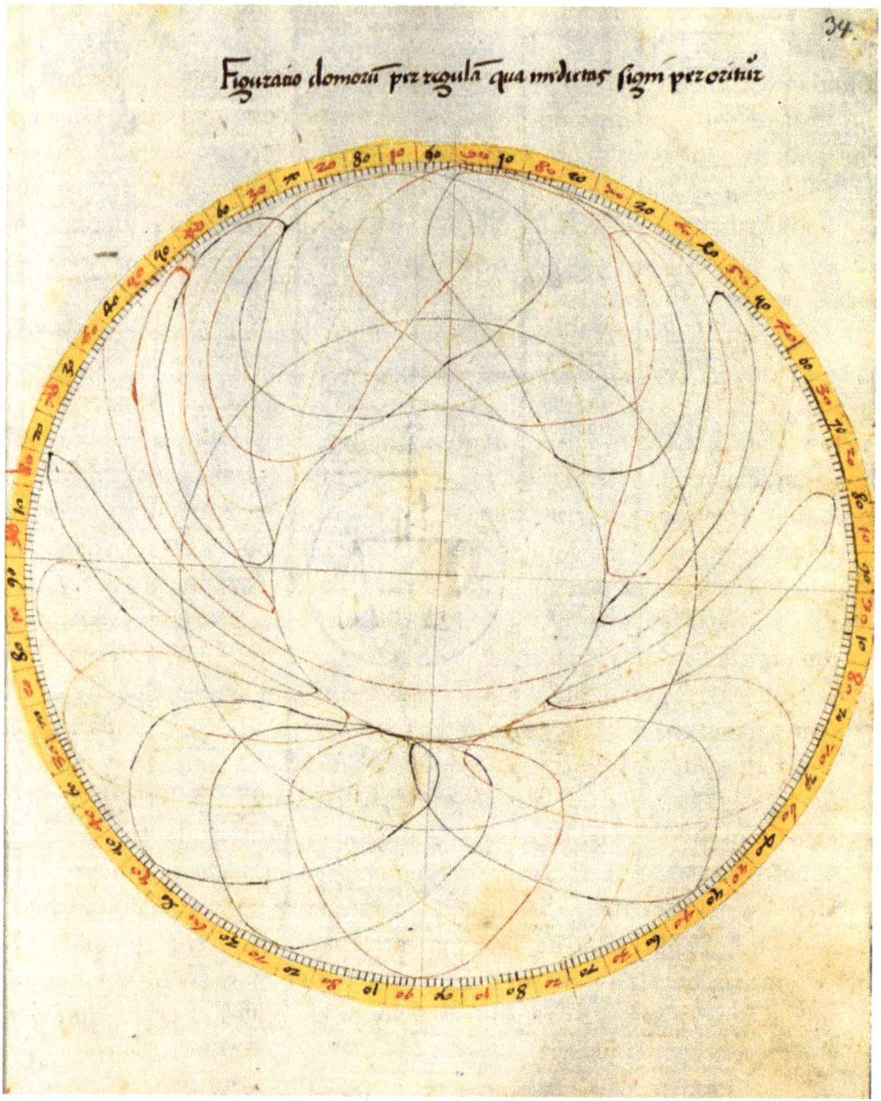

FIGURE 8.11 Ecliptic planetary hours for shortening and lengthening days on an astrolabe tympanum, anonymous drawing, *c.* 1520.
COURTESY BAYERISCHE STAATSBIBLIOTHEK, MUNICH, COD.ICON. 182, FOL. 34R.

hours, we should not be surprised that the Oppenheim and Görlitz sundials mark unequal hours of the first type.

I turn finally to the geometry of the astrological houses represented only on the Görlitz Arachne. In making a horoscope, the astrologer needs to know

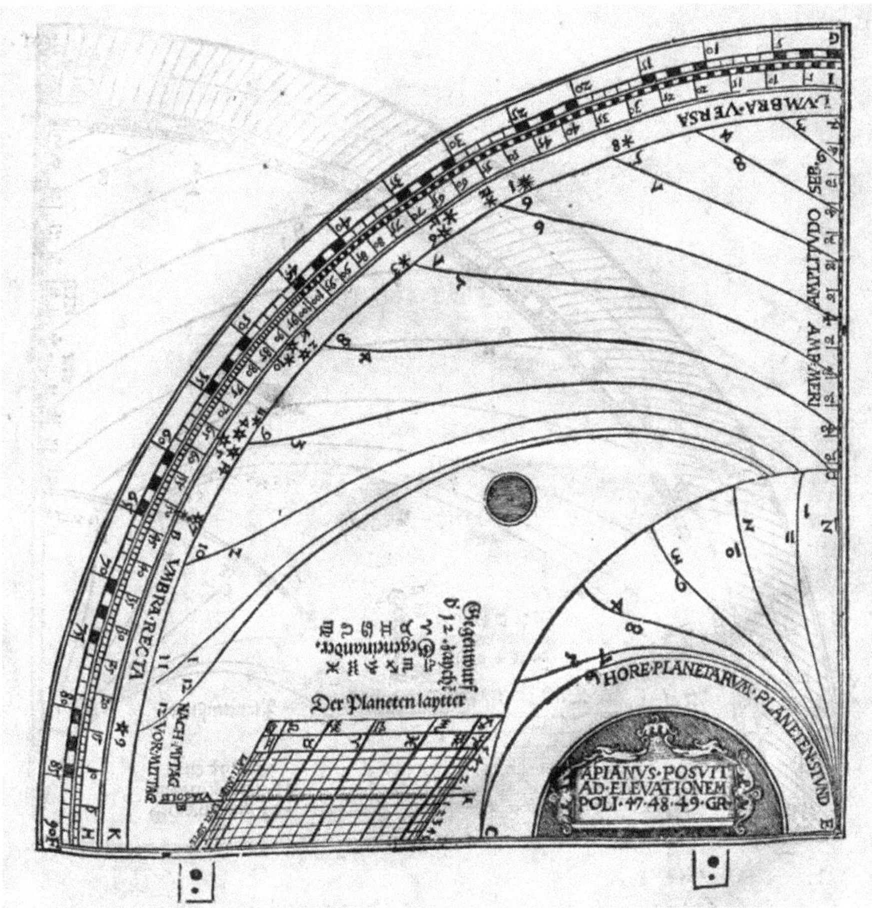

FIGURE 8.12 Ecliptic planetary hours represented on a portable quadrant-sundial; from Petrus Apianus, *Instrument Buch*, Ingolstadt, 1533, sig. a3v.
COURTESY BAYERISCHE STAATSBIBLIOTHEK, MUNICH, HBKS/R 2 GE.

the longitudes along the zodiac for each house boundary. Given knowledge of planetary longitudes, these bodies can be placed into their appropriate houses. On a sundial, however, one only reads the time during which the Sun remains in a given house; no separate knowledge of the solar longitude is needed. The house boundaries on the Görlitz dial, coupled with the Sun's shadow, would motivate a very narrow set of astrological predictions, i.e. those derived from the Sun's position in the astrological houses.

As John D. North has shown, at least seven different systems for dividing horoscope houses had emerged by the fifteenth century. In his *Tabulae directionum profectionumque*, printed in 1490 and many times thereafter,

FIGURE 8.13 An astrolabe-like instrument marking ecliptic planetary hours for lengthening days. In this copy of the book, the ecliptic volvelle and pointer for marking solar longitude have been erroneously attached, with their pivot point (the dark knot) not at the centre of the larger circle (tympanum). To represent the daily motion of the ecliptic against the local horizon this pivot should be at the centre of the tympanum; from Oronce Finé, *De duodecim caeli domiciliis et horis*, Paris, 1553, unpag.
COURTESY ÖSTERREICHISCHE NATIONALBIBLIOTHEK, VIENNA, 72.T.33* ALT.

Regiomontanus had described three of these systems, the most common or standard method often attributed to Alcabitius (tenth century), the prime vertical fixed-boundary method attributed to Campanus of Novara (*c.* 1210–1296) and Johann Gazulus (fifteenth century), and the equatorial fixed-boundary

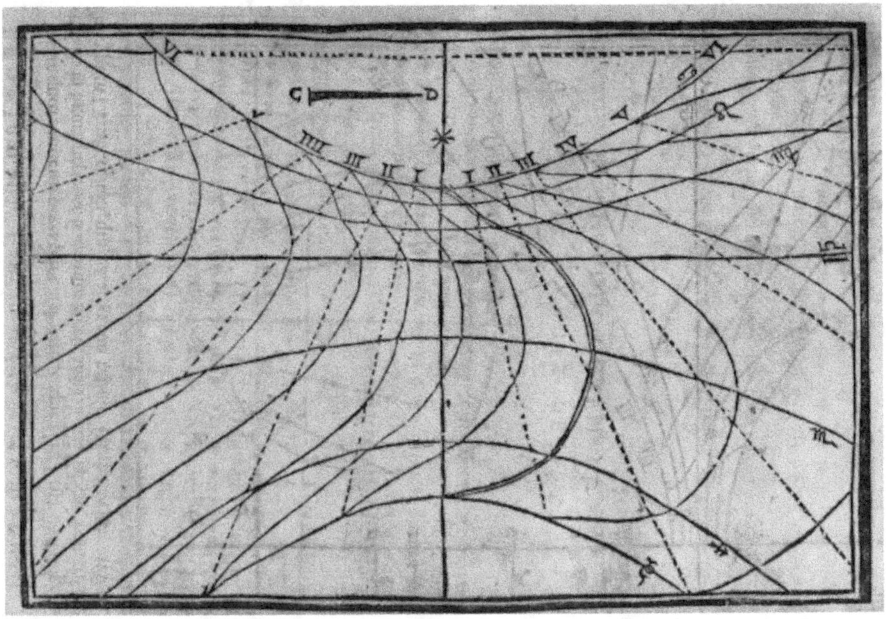

FIGURE 8.14 Ecliptic unequal hour lines drawn for a horizontal sundial; from Juan de Caramuel y Lobkowitz, *Solis et artis adulteria*, Louvain, 1644, p. 72.
COURTESY BAYERISCHE STAATSBIBLIOTHEK, MUNICH, 2 H.UN. 21.

method called "rational" and often attributed to Regiomontanus although now traced back to Ibn Mu'ādh (eleventh century).[38] Early modern astrologers stridently debated the merits of these systems. For example, Oronce Finé in his 1553 text announcing the new paper instrument combatively claimed that Campanus's method was more rational than Regiomontanus's (we need not rehearse the arguments here). Interestingly, Finé's twelve-page table, presenting the Campanus cusps to a precision of arcminutes (beyond what his paper instrument can yield), gives those values for only half of the houses; the other half have been computed erroneously.[39] Calculating cusps was a diffi-

38 John D. North, *Horoscopes and History*, London, Warburg Institute, 1986; Johannes Regiomontanus, *Tabulae directionum profectionumque*, Augsburg, Ratdolt, 1490, sig. B1r-B4r, O7v-O8r; E.S. Kennedy, "Ibn Mu'ādh on the Astrological Houses", *Zeitschrift für Geschichte der arabisch-islamischen Wissenschaften* 9 (1994), pp. 153–160.

39 Finé, *De duodecim caeli domiciliis*, cit. (n. 36). I have computed the Campanus and Regiomontanus cusps, as a function of the ascendant, following equations 11 and 12 offered by North, op. cit. (n. 38), p. 5. For a useful survey of sixteenth-century debates over house divisions, see Steven Vanden Broecke, *The Limits of Influence: Pico, Louvain, and the Crisis of Renaissance Astrology*, Leiden, Brill, 2003, pp. 236–251.

cult exercise in spherical trigonometry. As I have elsewhere argued, sixteenth-century mathematicians sought to invent easier means to find the house boundaries via instruments, sundials or tables of numbers.[40]

Fixed boundary methods lock the cusps to the local horizon. Lines for these boundaries can be drawn directly onto astrolabe tympani or sundial faces.[41] Lines for the Campanus and Regiomontanus methods vary significantly on vertical sundials. For the geographical latitude of Görlitz (51°), I compute the angle between the meridian line and lines for houses 9, 8 and 7 to be 30°, 60° and 90° for Campanus and 42.5°, 70° and 90° for Regiomontanus.[42] The lines of most recent renovation of the Görlitz Arachne precisely match values for the Regiomontanus boundaries; presumably the earlier versions made by Scultetus had also presented the Regiomontanus boundaries. The Görlitz dial thus matches the overwhelming majority of surviving sixteenth-century brass astrolabes in presenting Regiomontanus cusps for the horoscope maker.

Conclusion

Astrological markings on scientific instruments, in this case on astronomical clocks and vertical sundials, take us into the quotidian public spaces of early modern European towns. By 1600, inhabitants of nearly every town and city could read civil time, divided into 24 equal hours of the day, on mechanical clocks or sundials displayed on public buildings.[43] In a few cities with public astronomical clocks, they could also read astrological time, in unequal hours, on astrolabe tympani affixed to their dials. And in two German cities, Oppenheim and Görlitz, they could read astrological time on public sundials.

Astrological time, by the sixteenth century, clearly was not as widely available as was civil time. Perhaps the cost and complexity of building moving astrolabes into astronomical clocks retarded the spread of mechanical technologies for astrological time. Perhaps the lack of instructions in printed

40 Kremer, op. cit. (n. 1).

41 For illustrations of Campanus and Regiomontanus lines on astrolabe plates, see North, op. cit. (n. 38), pp. 64–65.

42 For Campanus's method, the house lines on a vertical dial are equivalent to the 30° segments of the prime vertical; for Regiomontanus's method that divides the equator into 30° segments, the house lines are equal to $\arctan(\tan M/\tan\phi)$, where $M = 0°, 30°, 60°$, etc. See Savoie, op. cit. (n. 26), pp. 126–127.

43 Zinner, op. cit. (n. 27), pp. vii–viii, recorded 5000 public sundials constructed before 1800 in more than 3000 towns; Ungerer, op. cit. (n. 21) in a less comprehensive survey found hundreds of public mechanical clocks. For a map illustrating the spread of astronomical clocks in Europe, see Eyraud, op. cit. (n. 21), p. 112.

gnomonic handbooks deterred dial makers from inscribing lines for planetary hours in their sundials. On the other hand, given the swelling flood of printed editions of annual astrological calendars and prognostications over the course of the sixteenth century, we probably should not conclude that public demand for astrological time had diminished in that century.[44] We shall leave it to social historians to investigate who purchased the printed ephemera and who patronized the making of public clocks and sundials. In any case, in Görlitz, civil and astrological time have been publicly available from 1550 to the present day, preserved through multiple restorations. And the sundials on the Ratsapotheke early became a public emblem for the city.

44 Klaus-Dieter Herbst (ed.), *Astronomie-Literatur-Volksaufklärung: Der Schreibkalender der Frühen Neuzeit mit seinen Text- und Bildbeigaben*, Jena, Verlag Historische Kalender Drucke, 2012.

CHAPTER 9

The Heavens at the Medici Court: Antonio Santucci's Cosmological Models

Giorgio Strano

The restoration between 2008 and 2010 of Antonio Santucci's large cosmological model preserved in Florence has brought to light information previously concealed beneath thick layers of dust and blackened varnish.* The information recovered helps to clarify the multifaceted purpose of the instrument.[1]

The design and construction occupied Santucci for years. Up to now, scholars have outlined three main steps: the making in 1582 of a first armillary, now at the Escorial, Madrid; the writing of the "Trattato sopra la nuova invenzione della sfera armillare" (Treatise on the New Invention of the Armillary Sphere) between 1582 and 1593; and, finally, the building of the Florentine device from 1588 to 1593.[2] Careful reading, however, has revealed that the "Trattato", preserved in a single manuscript copy, was in fact the first step of the process. Writing it clarified Santucci's thoughts on what he was trying to do, and convinced Ferdinand de' Medici, at the time a cardinal of the Holy Roman Church, to provide funding. Santucci's plan was realized twice, in Rome and Florence. In the items themselves, however, the functions of a versatile cosmographic compendium set out in the "Trattato" were partly changed.

---

* This chapter is an amalgam of presentations at the SIC symposiums in Boston (2007), Florence (2010) and Turin (2015). I would like to thank Karen Giacobassi and Karen Tomashavsky for their kind revision of the texts of the presentations, and Richard Dunn for that of this chapter.

1 Museo Galileo: Institute and Museum of the History of Science, Florence, inv. no. 714.

2 See Ferdinando Meucci, *La sfera armillare di Tolomeo costruita da Antonio Santucci*, Florence, Tipografia del Vocabolario, 1876, p. 13; Matteo Fiorini, *Sfere terrestri e celesti di autore italiano oppure fatte e conservate in Italia*, Rome, Società Geografica Italiana, 1899, p. 225; Maria Luisa Righini Bonelli, "Di alcuni manoscritti inediti di Antonio Santucci delle Ripomarance", *Annali dell'Istituto e Museo di Storia della Scienza di Firenze* 3, 2 (1978), pp. 59–67: 59–60, n. 2.

 | DOI:10.1163/9789004381438_011

The Obscure Life of a Court Cosmographer

Antonio Santucci was born in Ripomaranci (Pomarance), near Volterra, around the mid-sixteenth century. No information is available on his early life and studies, except that he was of humble origins.[3] As with many other astronomers, his biography must be reconstructed in part from observation reports.

Santucci was in the service of the Medicis from at least 1572, as one of the men Cosimo I placed around his son Ferdinand for education and guidance.[4] Ferdinand had become a cardinal in January 1563 and moved permanently to Rome from January 1569.[5] Santucci did not stay with Ferdinand at the beginning of his service. He observed the 1572 nova in Cassiopeia and the great comet of 1577 from Florence.[6] By March 1582, however, when another comet appeared, he was at Ferdinand's household, the Villa Medici of Trinità dei Monti, Rome.[7] He completed the "Trattato sopra la nuova inventione della sfera armillare" and his first armillary instrument the same year.

On the death of Ferdinand's brother, the Grand Duke Francesco, and Francescos's wife, Bianca Capello, in 1587, Ferdinand became Grand Duke of Tuscany, and Santucci moved to Florence as court cosmographer, keeping the title under Cosimo II. From 1588 to 1593, while teaching at the Accademia

3 Santucci's "passage from cobbler to reader of mathematics had been a great one [il transito da ciabattino all'esser lettore di matematica era stato un gran fare]"; Benedetto Castelli to Galileo Galilei, 10 December 1613, in Antonio Favaro (ed.), *Le Opere di Galileo Galilei: Edizione Nazionale sotto gli Auspicii di Sua Maestà il Re d'Italia*, Florence, Barbèra Editore, 1890–1909; reprint 1929–1939, 20 vols.: v. 11, p. 604 (if not otherwise stated, all translations from old Italian into English are mine). The term "ciabattino" (cobbler) may simply refer to a person of low standing and doing works of no meaning, such as repairs; see, for example, Benvenuto Cellini, *La vita, I trattati, I discorsi* (edited by Pietro Scarpellini), Rome, Gherardo Casini Editore, 1967, p. 342.

4 In the dedication to Cosimo II in Antonio Santucci, *Trattato nuovo delle comete*, Florence, Appresso Gio. Antonio Caneo, 1611, p. [1], he calls himself "old servant [servitor vecchio]" to the Grand Ducal house. He adds: "your Most Serene father Grand Duke Ferdinand (may he rest in the Heavens) gave me the occasion in Rome, and in Florence, to observe all the comets and new stars which have appeared during my life in 1577, 1582, 1596 and 1607, and the two stars of 1572 and 1604 [il Serenissimo Gran Duca Ferdinando suo Padre, (che sia in Cielo) mi dette occasione in Roma, et in Firenze di poter osservare tutte le comete, e nuove stelle apparse al mio tempo del 1577, del 82, del 96, e del 607, et le due stelle del 72, e del 1604]", *ibid.*, pp. [1–2].

5 Elena Fasano Guarini, "Ferdinando I de' Medici", in *Dizionario Biografico degli Italiani*, Rome, Treccani, 1960-, 91 vols. (to date): v. 46 (1996), pp. 258–278: 260.

6 Santucci, op. cit. (n. 4), pp. 30 and 55–57.

7 *Ibid.*, p. 67.

Medicea del Disegno,[8] he supervised the construction of another large armillary for the Maps Room of the Uffizi Gallery. In 1590, he also built a quadrant with a "diameter" of 4.5 Florentine *braccia* (*c.* 2.6 m), divided to one arc minute.[9] In 1593–1594, Ferdinand commissioned him to prepare a beautifully illustrated "Trattato di diversi istrumenti matematici" (Treatise on Several Mathematical Instruments), on the most remarkable items in the Wardrobe of the Palazzo Vecchio.[10] In 1595, he also refurbished the large terrestrial globe by Egnazio Danti (1536–1586), moved from the Palazzo Vecchio to the Uffizi Gallery.[11] In July 1596, he observed a new comet from the royal Pitti Palace.[12]

Following a formal request to the Grand Duke for permanent employment at court, Santucci received a small salary and, in 1599, succeeded to Francesco Sanleonini as reader of mathematics at the University of Pisa, where he taught until his death.[13] In his new position at Pisa, he observed the nova in Sagittarius in 1604–1605, using his quadrant and another brass instrument of a half-*braccio* (*c.* 29 cm) in diameter.[14] In 1606, he presented the Grand Duke with twelve geographical maps, on which he had been working since 1600.[15] Finally, he observed the comet of October 1607 from Pisa.[16]

8 The book of the Provveditori of the Academy for 1591–1594 includes an inventory by Jacopo di Chimenti of Empoli, dated 28 June 1591, listing "two shelves attachable to the desk, in order to place on them the spheres, when magister Antonio of Pomarance is lecturing [dua mensole da attachare alla chattedra per posarvi le palle sferiche quando legge m.o Antonio dalle Pomarance]": Florence, Archivio di Stato di Firenze (ASF), Acc. Dis. 27, fols. 131v–133v; transcribed in Zygmut Waźbiński, *L'Accademia del Disegno a Firenze nel Cinquecento*, Florence, Olschki, 1987, p. 487.

9 Santucci, op. cit. (n. 4), p. 83.

10 Santucci, "Trattato di diversi istrumenti matematici che si conservano al presente nella Guardaroba del Gran Duca di Toschana [...]", 1593–1594: Florence, Biblioteca Marucelliana, Ms. C 82. Composition dates are mentioned on fol. 7v: "in this present year 1593 [in questo presente anno 1593]", and fol. 66v: "this year 1594 [questo anno 1594]". See Marta Bertini, "Il *Trattato di diversi istrumenti matematici* di Antonio Santucci", *Nuncius* 17, 1 (2002), pp. 247–262.

11 Meucci, op. cit. (n. 2), p. 11.

12 Santucci, *Trattato nuovo delle comete*, cit. (n. 4), p. 75. Santucci wrongly dates the comet to 1597.

13 Meucci, op. cit. (n. 2), p. 12; Favaro (ed.), op. cit. (n. 3): v. 20, p. 532. Santucci's position is confirmed in two letters: Costanzo da Cascio to Galileo, 24 May 1604: *ibid.*, v. 10, p. 108; Giovanni Talentone to Galileo, 18 June 1612: *ibid.*, v. 11, p. 332.

14 Santucci, *Trattato nuovo delle comete*, cit. (n. 4), pp. 83–84.

15 Meucci, op. cit. (n. 2), p. 12. A thirteenth map was still in preparation.

16 Santucci, *Trattato nuovo delle comete*, cit. (n. 4), pp. 95–96 and 99.

Around 1609–1610, Santucci wrote a short manuscript treatise on cosmography.[17] He also published the *Trattato nuovo delle comete* (New Treatise on Comets, Florence, 1611), with a second edition appearing posthumously in 1619. Based on parallax measurements, the treatise supported the theory that the new stars and comets were in the celestial regions, not the sublunary regions of fire and air. In 1612, Santucci wrote the "Breve discorso sopra il trattato del Sig. Galileo Galilei delle cose che galleggiano sopra l'acqua" (Brief Discourse on Galileo Galilei's Treatise on Bodies Floating on Water), which, once again, remained unprinted.[18] He died in 1613.[19]

The Project of the "New Armillary Sphere"

Armillary spheres are representations of the celestial sphere. They include a polar axis, which may pass through a small terrestrial globe, and eight rings for the celestial equator, ecliptic, tropics of Cancer and Capricorn, Arctic and Antarctic polar circles, and colures for the equinoxes and solstices (*Fig. 9.1*). They were either held in the hand or placed on a stand, in which case there are at least two additional rings: local meridian and horizon. Any modification to this structure makes an armillary sphere into another device.

The stimulus to build large armillaries—that is, instruments made of rings (Latin *armilla* = armlet)—probably came from Ferdinand. The future Grand Duke had an interest in astronomy and astrology, apparent in Iacopo Zucchi's frescos painted between 1572 and 1587 in the cardinal's Villa Firenze and Villa

17 Santucci, ["Trattato di cosmografia"], *c.* 1611: Florence, Biblioteca Nazionale Centrale di Firenze (BNCF), ms. Magliabechiano XIII, 17. The composition date comes from fols. 5r–5v. Pedro Sarmiento is said to have passed the Magellan Strait "30 years ago", referring to a voyage in 1579–1580.

18 Again, the composition date is from the manuscript: Santucci, "Breve discorso sopra il trattato del Sig. Galileo Galilei delle cose, che galleggiano sopra l'acqua [...]", *c.* 1612: Florence, BNCF, Ms. Gal. 316, cc. 447r-[458v]: [457v]: "in this Winter of the years 1611 and 1612 [in questa invernata dell'anno 1611 et 1612]". See Francesco Paolo de Ceglia, *De natantibus: Una disputa tra filosofia e matematica nella Toscana medicea (1611–1615)*, Bari, Laterza, 1999, pp. 156–163.

19 Benedetto Castelli to Galileo, 10 December 1613; in Favaro (ed.), op. cit. (n. 3), v. 11, p. 604. On Santucci's life, see Righini Bonelli, op. cit. (n. 2), pp. 59–60, n. 2; Thomas B. Settle, "Antonio Santucci, his *New Tractatus on Comets*, and Galileo", in Paolo Galluzzi (ed.), *Novità celesti e crisi del sapere: Atti del convegno internazionale di studi galileiani*, Florence, Giunti, 1983, pp. 229–238: 231–234; F.P. de Ceglia, "Santucci, Antonio, delle Pomarance (Ripomarance)", in *Dizionario Biografico degli Italiani*, cit. (n. 5): v. 90 (2017), pp. 479–482.

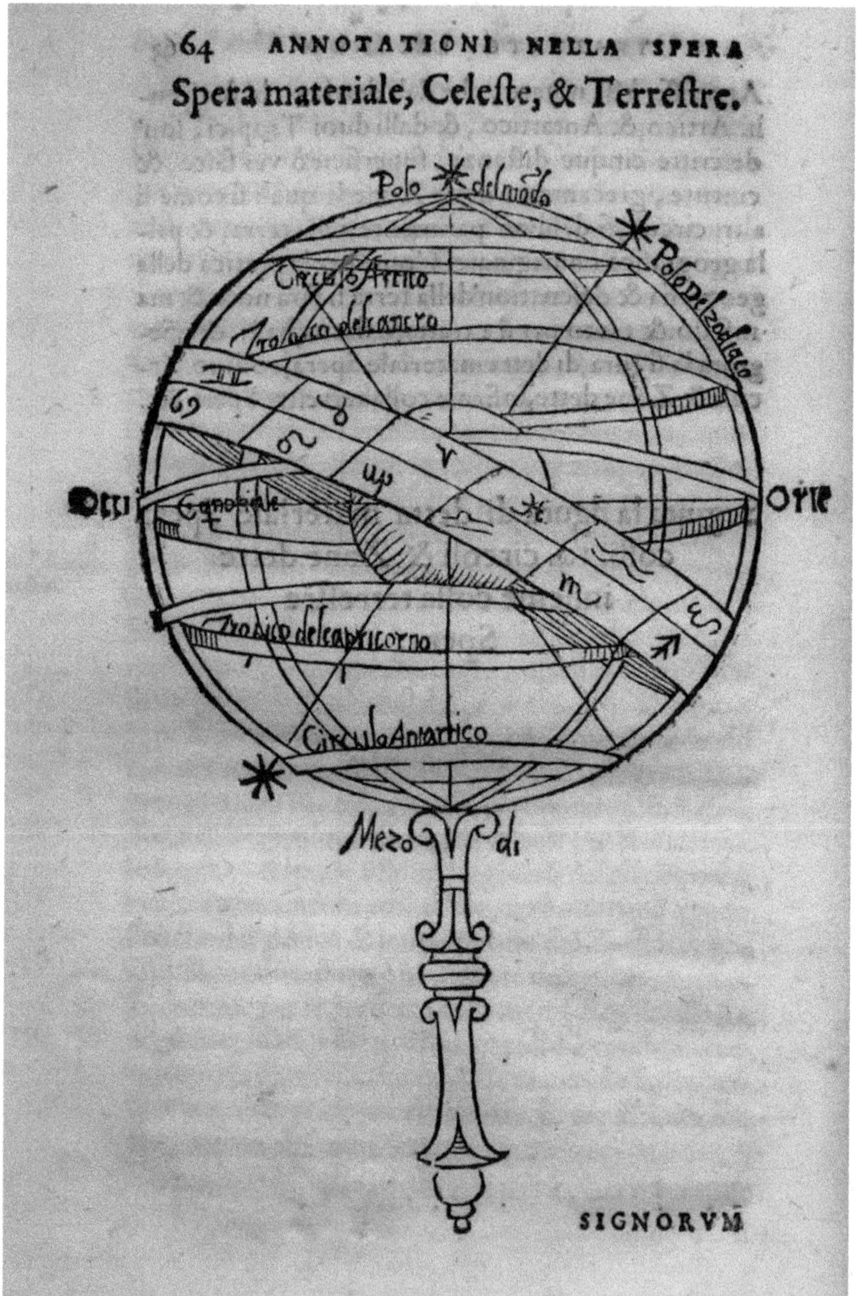

FIGURE 9.1 The basic structure of the armillary sphere; from Mauro da Firenze, *Annotationi sopra la lettione della Spera del Sacro Bosco*, Florence, Lorenzo Torrentino, 1550, p. 64.

Medici, in Rome. With this encouragement, Santucci finished his "Trattato" on a new type of armillary sphere in early 1582.

The composition date emerges from the closing "Advice". All the effort expended on the Gregorian calendar, it notes, will have been worthwhile "when the above mentioned reform comes into use".[20] Unless Santucci got informal news concerning the new calendar, this would date the "Advice" to between Pope Gregory XIII's papal bull *Inter gravissimas* (24 February 1582) and its enforcement from 4 October 1582. The rest of the manuscript, which includes examples using the Julian calendar, predates the bull.

The prose of the "Trattato" is cumbersome. Structural and procedural details are described more than once and redundantly. Like Santucci's other works, the language is not Latin, but Italian. The writing displays erratic punctuation, random capitalization and variant spellings, even of the same word on the same page. Hypercorrections are frequent, involving the excessive use of "h", mute in Italian. Spaces between words are misplaced and follow patterns of speech rather than written forms. These characteristics confirm that the author was not from a learned family but was an artist-craftsman with little formal education, as was quite common in the Italian states at that time.[21]

Following a traditional structure, the "Trattato" is divided into two parts on the making and use of the instrument. In particular, Santucci might have had at hand Danti's treatise *La sfera di Proclo* (Proclus's sphere), printed in 1573. In addition to the commented translation of the writing on the celestial sphere by Proclus Lycaeus Diadochus (412–485), that work contains Danti's original *Trattato sull'uso della sfera* (Treatise on the use of the sphere).[22] Santucci's "Trattato" partially reflects the bipartite structure and the order of chapters of *La sfera di Proclo*. It is therefore very likely that Santucci's "new armillary sphere" was conceived as the implementation of the traditional armillary sphere described by Danti.

In the first part of Santucci's treatise, the first chapter, of twenty, is a compilation of basic cosmography, unrelated to a specific instrument. It outlines the cosmos as a compound of ten concentric spheres: seven for the planets, one

20 Santucci, "Trattato sopra la nuova inventione della sfera armillare ad istantia dell'Illustrissimo et Reverendissimo Monsignore Don Ferdinando Cardinal de' Medici composto da Antonio Santucci da Ripamaranci inventore di detto istrumento suo devotissimo vassallo", *c.* 1582: Genoa, Biblioteca Universitaria, Ms. F.VII.6, fol. 64v.

21 Settle, op. cit. (n. 19), pp. 231 and 234; De Ceglia, op. cit. (n. 18), p. 157 and n. 36.

22 Egnazio Danti, *La sfera di Proclo Liceo* [...], Florence, nella Stamperia de' Giunti, 1573. The annotated translation occupies pp. 1–55, the *Trattato sull'uso della sfera* follows a new numbering, pp. 1–33.

each for the stars, first mover and crystalline sky.[23] Chapters 2 to 8 describe the ten circles of the armillary sphere with stand. Chapters 9 to 14 describe the additional components: a revolving vertical circle for altitudes; five fixed parallel circles corresponding to those of the celestial sphere (equator, tropics and polar circles), four position circles, 24 circles for the seasonal hours, twelve meridian circles, and an horary dial with three hands. Chapter 14, on the horary dial, includes the "Theory of the Sun", or solar calendar, to be placed on the horizon ring. Chapters 15 to 19 deal with the geography of the terrestrial globe, while the final chapter 20 recapitulates the whole structure of the armillary. There is not a single word on the size of the sphere or materials used.

The second part, 40 chapters on the instrument's use, mostly deals with the operations common to a planispheric astrolabe. It also reveals Santucci's ambitious plans. Occasionally, previously unmentioned construction details emerge in the discussion of particular operations. Their displacement to the second section reveals that the treatise was a work in progress. New ideas on the use of the armillary were conceived during the writing and needed ad hoc structural solutions.

Observation: The armillary is placed outside and aligned using the magnetic compass in the base.[24] The polar axis is hollow and works as a sighting tube.[25] By rotating the sphere horizontally and vertically, it is possible to aim the axis at the Sun, planets or stars to measure their azimuthal co-ordinates.

---

23 Like other Tuscan astronomers, Santucci accepted the theory of the trepidation of the equinoxes. The mechanical explanation of the theory required the additional sphere of the crystalline sky. The trepidation mechanism appears, for example, in the largest of the four "Theories of the planets" completed between 1570 and 1574 by Antonio Lupicini (*c.* 1530–*c.* 1598) for the Biblioteca Medicea Laurenziana, Florence. Elly Dekker, Giorgio Strano, Carlo Triarico, *Teoriche dei pianeti e teorie dell'Universo*, Florence, Istituto e Museo di Storia della Scienza, 2000, no. 1; Elly Dekker, *Istituto e Museo di Storia della Scienza. Catalogue of Orbs, Spheres and Globes*, Florence, Giunti, 2004, pp. 34–39; Mara Miniati, "'Un fabbro che sia buon maestro': La produzione di strumenti scientifici a Firenze nel Cinquecento", in Maurizio Bossi, Gloria Fossi *et al.* (eds.), *Arti Fiorentine: La grande storia dell'artigianato*, Florence, Giunti, 1998–2003, 6 vols.: v. 3 "Il Cinquecento" (2000), pp. 263–295: 280 and nn. 31, 32.

24 The magnetic compass is mentioned briefly in the first part of Santucci, "Trattato sopra la nuova inventione", cit. (n. 20), fol. 20v, with a longer explanation in the second, *ibid.*, fol. 29v.

25 In the first part of the "Trattato", the axis is mainly discussed as a theoretical entity. The only material reference is in ch. 14, where the axis holds the three hands of the horary dial; *ibid.*, fol. 19v. In the second section, Santucci unexpectedly claims, "the axis of the sphere being hollow, in the shape of a tube, as already said [essendo l'asse della sfera perforata, come s'è detto, a guisa d'una cannella]"; *ibid.*, fol. 31v.

Timekeeping: By aiming the hollow axis at the Sun or a star, the armillary performs the timekeeping functions of a planispheric astrolabe.[26] To that end, the sphere is equipped with a Sun pointer that slides along a lute string embedded along the ecliptic band,[27] and has a number of star pointers connected to the rings of the eighth sphere.[28] The altitude circle helps to position the pointers from the altitude measurements.

The armillary also contains twelve fixed meridian circles and another twelve circles forming a sphere with a horizontal north-south axis. These rings allow the user to tell the time according to five systems: Italian hours (from sunset);[29] Bohemian hours (from sunrise);[30] astronomical hours (from noon);[31] Spanish, French or German hours (from midnight);[32] and unequal, seasonal or planetary hours (twelve hours by day, twelve by night).[33]

Astronomical computations: The position circles and pointers allow the user to perform other calculations for which astrolabes were generally used. These are described in quite closely grouped chapters: the Sun's position on the zodiac, meridian and horizon (chapters 2–4, 6); horary conversions (13, 14, 17, 18, 22); time and duration of solar (23–26), stellar (23–26, 28–32) and lunar (33–34) phenomena. Chapter 39 deals with the prediction of tides, a topic of particular interest to the Medicis, eager for maritime control.

Education: The classic use of armillary spheres finds an unusual variant in chapter 27: "How to identify the planets in the sky, and all the stars of the sphere, without knowing any of them". Again, the armillary is placed outside,

26 For the use of the armillary by day and night, see *ibid.*, fols. 32v–33r.

27 "[...] the ecliptic line must be slightly grooved, in order to host a lute string, on which the Sun and the other planets can be placed [la linea eclitica sia cavata sottilmente; tantoché vi s'accomodi drento una corda da leuto, nella quale vi si raccomoda il Sole, et gl'altri pianeti]"; *ibid.*, fol. 20v.

28 "[...] the sphere must contain thirty or forty of the brightest stars of the sky, adapted to the intrinsic circles and steadily attached to them by small arms, as explained by Ptolemy in the seventh book of the *Almagest* [bisogna che nella sfera sieno descritte trenta, o quaranta stelle, delle più ilustri del cielo, le quali s'accomodano a circoli intrinsici, che stieno attaccate a quelli con certi razzetti immobilmente in quel modo, che Tolomeo insegna nel settimo libro del Almagesto]"; *ibid.*, fols. 20r–20v. Claudius Ptolemy is perhaps mentioned for his star catalogue: *Almagest*, VII, 5 and VIII, 1.

29 Santucci, "Trattato sopra la nuova inventione", cit. (n. 20), fols. 32v–33r.

30 *Ibid.*, fol. 33v.

31 *Ibid.*, fols. 33v–34v.

32 *Ibid.*, fols. 34v–35r.

33 *Ibid.*, fols. 39r–39v. The set of rings with horizontal poles marks the unequal hours very roughly. The correct layout of the unequal hour lines is explained in a treatise that would have been available to Santucci: Egnazio Danti, *Trattato dell'uso et della fabbrica dell'astrolabio*, Florence, Giunti, 1569, pp. 33–35.

aligned north-south and set for the observer's latitude and the time. With the star pointers in position, the planets are set along the lute string in the middle of the zodiac using ephemerides for that date.[34] The user can then compare the armillary and the sky.

Astrology: According to the first part of the "Trattato", the armillary is an astrological compendium. The zodiac ground colours reflect the element of each sign: red for fire; ochre for earth; blue for air; green for water.[35] The zodiac also includes the "qualities" of the signs: masculine or feminine, hot or cold, dry or wet.[36] With these sparse tips, the armillary works as an astrological calculator. Five chapters of the second part (21, 35–38) discuss how to find the planet ruling each hour, the boundaries of the twelve celestial houses, drawing nativity charts, calculating the revolutions of the years, and the directions of the planets and key stars.

The ruling planet depends on the unequal hour circles. The descending sequence Saturn, Jupiter, Mars, Sun, Venus, Mercury, Moon, is set on the planet ruling the day. For example, on Monday, Moon's day, the first hour after sunset is ruled by the Moon. By progressing along the planetary sequence, Jupiter rules Monday's last hour. The next planet, Mars, rules both the new day, Tuesday, and its first hour.[37] The relevant table of the "Trattato" would suffice for any calculation.[38] Nevertheless, Santucci claims that the armillary simplifies the work with an ad hoc device: a "planetary ring" divided into 24 equal arcs that slides above the celestial equator. When the arc of the planet of the first hour is set in place, the other planetary arcs align with the correct hours.[39]

By skipping the even circles of the unequal hours, the remaining six are the boundaries of the twelve celestial houses. Santucci devotes a chapter to explaining that the houses start below the eastern horizon and, by looking at the sphere from the north, are numbered anticlockwise from one to twelve.[40] Once the planetary pointers are set along the ecliptic with the ephemerides, and

34 Santucci, "Trattato sopra la nuova inventione", cit. (n. 20), fols. 45r–46r.

35 *Ibid.*, fol. 8v. The "green" for water is incorrectly given as "red", which is used for fire.

36 *Ibid.*, fols. 8v–9v.

37 *Ibid.*, fols. 39v–40r.

38 *Ibid.*, fol. 40v.

39 *Ibid.*, fol. 41r. Santucci neglects the fact that 24 is not divisible by 7, which prevents a simple division of the planetary ring; and that, as well as the incorrect placement of the unequal hours circles, equal divisions of a circle parallel to the equator do not match equal wedges of a sphere with a horizontal axis.

40 *Ibid.*, fols. 53v–54v.

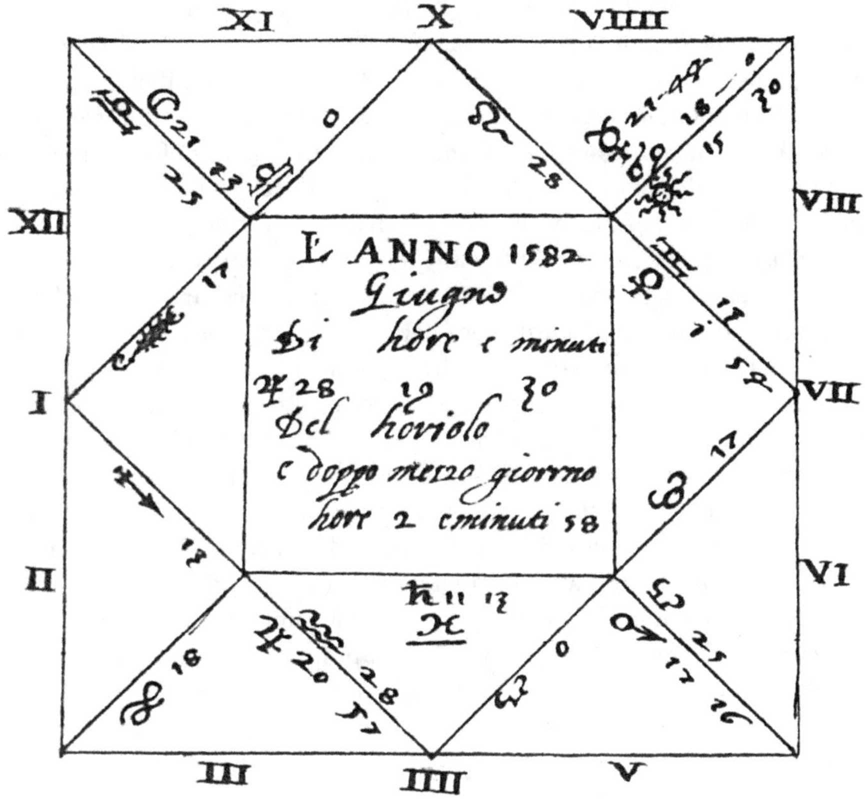

FIGURE 9.2 The positions of the zodiacal signs and planets in the twelve houses for 28 June (Thursday) 1582 (Julian calendar) at 19:30 Italian hours or 2:58 astronomical hours; from Antonio Santucci, "Trattato sopra la nuova invenzione della sfera armillare", *c.* 1582 (Genoa, Biblioteca Universitaria, MS. F.VII.6, f. 56v).

the armillary is rotated for latitude and time, the degree of the zodiac beginning each house and the position of the planets in them are easily read off (*Fig. 9.2*).[41]

The next step in prognostication consists in determining the revolutions of the years for nativities, for which Santucci refers to Alfonso X, King of Castile

41 *Ibid.*, fols. 54v–56v. On Renaissance astrological diagrams, see Giuseppe Bezza, "Representation of the Skies and the Astrological Chart", in Brendan Dooley (ed.), *A Companion to Astrology in the Renaissance*, Leiden, Brill, 2014, pp. 59–86: 59–65. Of the different methods for establishing the houses, Santucci selects the one associated with the planispheric astrolabe; *ibid*, p. 65, and John D. North, *Horoscopes and History*, London, Warburg Institute, 1986, pp. 20–27 and 46. This method was usually ascribed to Campanus of Novara (*c.* 1220–1296), *ibid.*, pp. 27–30.

and Leon (1221–1284), Giovanni Bianchini (1410–1469) and Hans Müller of Königsberg (Regiomontanus, 1436–1476).[42] To draw the celestial chart, the armillary user must take note of the degree in which the local meridian cuts the celestial equator, then turn the armillary clockwise 87° 18′ for every year since birth.[43] Once the armillary is rotated, and the planets placed along the ecliptic for the day of the completed revolutions, the chart for the new year—called the "diagram of the revolutions"—can be drawn.[44]

The last step is determining the directions of the planets and other stars. Santucci mentions a single reference, Ptolemy (second century AD), in relation to the predictive formula "one degree one year".[45] The two planets of which the user wants to know the direction—the "significator" (the one that signifies) and "promittor" (the one that promises)—are placed on the ecliptic according to the ephemerides for that day. The user can then convert ecliptic longitudes into right ascensions, and determine the angular separation of the planets.[46]

Nowhere does the "Trattato" explain the principles of prognostication. With the nativity chart, revolutions of the years and planetary directions, the future court cosmographer was apparently relying on the principle that the life events of an individual, or anything for which a commencement date can be established, are predictable by drawing the appropriate chart and applying the principle "one degree one year" to the planetary directions. The aspects and influences of the zodiacal signs, mentioned in the first part of the "Trattato",[47]

42 Santucci, "Trattato sopra la nuova inventione", cit. (n. 20), fol. 57r. The references are presumably to the "Alphonsine Tables", Bianchini's *Tabulae astronomiae* (1495) and Regiomontanus's *Tabulae directionum* (1475).

43 The angle of 87° 18′ is the difference in right ascension between the tropic year of 365 d, 5 h, 49 m, 16 s (from the "Alphonsine Tables") and the Julian year of 365 days; *ibid.*, fols. 56v–57r. Several transcription errors on fol. 57r suggest that the manuscript is a late draft. The tropic year is miswritten 365 d, 5 h, 45 m, 16 s. The difference of 87° 18′, based on the correct tropic year, becomes 87° 49′, and 250° + 87° makes 237° instead of 337°.

44 *Ibid.*, fols. 57r–57v.

45 *Ibid.*, fol. 58r. The formula appears in Claudius Ptolemy, *Tetrabiblos*, III, 11, 16, to calculate a person's lifespan. See the bilingual Greek-Italian edition: C. Ptolemy, *Le previsioni astrologiche* (*Tetrabiblos*), trans. by Simonetta Feraboli, Milan, Mondadori, 1985, pp. 227–229, and remarks on pp. 435–436. The formula reappears in *Tetrabiblos*, IV, 10, 19. The chapter numbers within each book of the *Tetrabiblos* have been rearranged with respect to the classic bilingual Greek-English edition: C. Ptolemy, *Tetrabiblos*, ed. and trans. by Frank Egleston Robbins, "Loeb Classical Library 435", Cambridge MA, Harvard University Press, 1940 (reprint: London, William Heinemann Ltd., 1980). Therein the formula appears in III, 10, p. 289, and IV, 10, p. 451. In the following, references to the Loeb edition will be added within square brackets.

46 Santucci, "Trattato sopra la nuova inventione", cit. (n. 20), fols. 57v–58v.

47 About which see Ptolemy, *Tetrabiblos*, I, 12, 1 to I, 20, 7 [I, 11 to I, 19, pp. 65–91].

also come from Ptolemy.[48] Santucci was no innovator and followed the trend for recovering Ptolemaic notions.[49] Nevertheless, a few elements of Medieval astrology—especially those gathered from Abū Maʿshar al-Balkhī (787–886)—remain alongside Ptolemy's. The revolutions of the years was one of these.[50]

The Madrid Armillary

The connections between the "Trattato" and the instruments are somewhat puzzling. Santucci's first armillary, less than two metres high, was finished in Rome before 21 August 1582, when Ferdinand ordered it shipped to Giulio Battaglini in Naples, an agent of Pietro de' Medici in Spain. Through the mediation of Cardinal Claudio La Baume, Viceroy of Naples, and the sphere's dedicatee, Ferdinand sent the armillary as a gift to King Philip II of Spain.[51]

The gift carried expectations. Ferdinand supported a pro-Spanish policy. In addition to his mediation within the Papal court to renew fiscal exemptions for Spain, the gift was intended to allay Philip's suspicions about the loyalty of the Medici house. The mediation and the gift were rewarded in early 1583, when Ferdinand received the prestigious title of "Protector of Spanish Affairs".[52]

The armillary left Naples by sea in February 1583 and, making port in Alicante, arrived in Madrid in June. As Battaglini wrote to Pietro Usimbardi, Ferdinand's secretary, the sphere arrived in a poor state: "the gilt and colours have mostly come off, most of the stars have fallen off and many circles and joints have come unstuck. I am now intending to have it restored with all care and great expense".[53] The restoration proceeded quickly. Philip II received the item by the end of 1583. He kept it at the Monastery of the Escorial in Madrid, where it remains in the main library (*Fig. 9.3*).[54]

---

48 See, in particular, *ibid.*, IV, 10, 13–27 [IV, 10, pp. 447–459].

49 Ornella Faracovi, "The Return to Ptolemy", in Dooley (ed.), op. cit. (n. 42), pp. 87–98: 92–98.

50 Bezza, op. cit. (n. 42), pp. 82–86.

51 Righini Bonelli, "The Armillary Sphere in the Library of the Escorial in Madrid", *Vistas in Astronomy* 9 (1968), pp. 35–40: 37–39. The dedication to La Baume appears on the cartouche of the terrestrial globe.

52 Fasano Guarini, op. cit. (n. 5), pp. 262–263.

53 Giulio Battaglini to Pietro Usimbardi, 4 July 1583: Florence, ASF, *Mediceo del Principato*, filza 5112, fol. 546v.

54 Righini Bonelli, "The Armillary Sphere", cit. (n. 51), pp. 37–40; Righini Bonelli, "Di alcuni manoscritti inediti", cit. (n. 2), pp. 59–60, n. 2; Miniati, op. cit. (n. 23), p. 280.

FIGURE 9.3 Armillary sphere, by Antonio Santucci, Rome, 1582 (Main Library of the Escorial, Madrid; Photo: Filippo Camerota).

This armillary only partially matches the description of the "Trattato". Not every difference is attributable to the damaging voyage and subsequent restoration.

Observation: The armillary has no observing function. There is no magnetic compass, so the instrument cannot be aligned north-south. The Earth axis is a brass tube, but it is unclear whether it can function as a sighting tube. The meridian ring fits in a notch of the horizon, preventing the taking of azimuths.

Timekeeping: The inability to use the armillary for observation implies that it cannot tell the time either. It can, however, perform horary conversions in the five systems mentioned by Santucci. In particular, it includes the horary dial and full set of rings for unequal hours (although these are incorrectly positioned in the present display).

Astronomical computation: Most of the calculations described in the "Trattato" can be performed. The armillary originally had star pointers, which were lost during its move to Madrid. The upper face of the horizon ring contains the theory of the Sun, essential for determining solar longitude. At the centre of the zodiac band, the grooved ecliptic could hold a lute string for planetary pointers. The zodiac also has a second groove, barely mentioned in the "Trattato",[55] tilted at about 5° over the ecliptic, which it intersects at the equinoxes. This second groove—for another lute string—represents the Moon's orbit (ignoring the progression of its nodes) and can be used for calculations involving lunar latitudes (*Fig. 9.4a*). In addition, the Madrid armillary contains a large ring with fourteen scales (currently mounted upside-down; *Fig. 9.4b*). The "Ruota perpetua" (perpetual wheel) that Santucci prepared later, after 1589 (*Fig. 9.5*),[56] indicates that this device helps in finding the dates of new Moon within the Metonic cycle. The presence of the second groove and this ring suggests that, beyond the structure outlined in the "Trattato", the Madrid armillary can perform calculations for the first appearance of the Moon after its conjunction with the Sun, and for the tides' state. Also not described in the "Trattato", it can perform calculations of solar and lunar eclipses. The same additional ring, however, highlights the fact that the "Trattato" does not specify what lies in the space between the terrestrial globe and the sphere of the stars. As Santucci writes that the total number of rings of the armillary is the usual ten, plus 34 for its additional functions (one altitude circle, five extrinsic parallels, four position circles, twelve unequal hour circles and twelve meridian circles), this space should be empty.[57]

---

55 Santucci, "Trattato sopra la nuova inventione", cit. (n. 20), fol. 31r.

56 Santucci, "Ruota perpetua nuovamente ad utilità comune posta in luce [...] nella quale perpetuamente si trova l'hora del levar del Sole, del mezo giorno, il far della Luna e tutti gli aspetti che fa, feste mobili, lettera dominicale con bisesto, aureo numero, epatta, ciclo solare, et indicione, cominciando dalla nuova riforma", Florence, after 1589: Museo Galileo: Institute and Museum of the History of Science, Florence, inv. no. 3716.

57 Santucci, "Trattato sopra la nuova inventione", cit. (n. 20), fol. 28v. Santucci's summary: "circle of altitudes [...], five extrinsic parallels [...], twelve position circles [...], twelve meridians and two theoretical circles", makes a total of 32, not 34. The difference comes

FIGURE 9.4 Armillary sphere, by Antonio Santucci, Rome, 1582; *a*) detail of the zodiac band with the two grooves for the ecliptic and lunar orbit; *b*) detail of the ring for calculating the date of the new Moon (Main Library of the Escorial, Madrid; Photos: Filippo Camerota).

FIGURE 9.5 Antonio Santucci, "Ruota perpetua nuovamente ad utilità comune posta in luce", Florence, after 1589 (Museo Galileo: Institue and Museum of the History of Science, Florence, inv. no. 3716).

Education: The missing compass aside, the Madrid armillary could have been used for identifying stars and planets.

Astrology: The Madrid armillary is a modest compendium. The ground colour identifies the element of each zodiacal sign and a few symbols mark the increased "influence" of a planet transiting specific regions. There is no indication of the signs' qualities. The armillary can, however, perform some of the astrological calculations described in the "Trattato". On the one hand, the circles of the celestial houses are distinguished from the additional circles that complete the sequence for the unequal hours. According to several drawings in the "Trattato", the circles of the houses are coloured, large, and divided in degrees; the additional circles of the hours are gilt, thin and undivided. On the other hand, the ad hoc additional device to determine the ruling planets is missing. The fixed celestial equator of the sphere is a single ring, divided on each edge.

The Madrid armillary represents both an involution and an evolution of the hypothetical instrument of the "Trattato". Such opposing tendencies probably result from the play-off between Santucci's theoretical ambitions, the limits of his skills and Ferdinand's moderate funding.

The Florence Armillary

On 4 March 1588, on the instructions of Ferdinand I, the new Grand Duke of Tuscany, Santucci began the construction of a second large armillary.[58] Preparations took until 6 May 1593.[59] The new court cosmographer coordinated many craftsmen to make wood and brass rings, nails, wood carvings, gilding, etc. The painting, formerly attributed to the school of Giorgio Vasari,[60] was executed by the cosmographer's brother, Enea Santucci.[61] This collaborative work is revealed by lists of materials, requests for payment,[62] and a report that a diligent clerk prepared for Ferdinand. This lists all the expenses for a total of 1052.2 Tuscan *scudi*.[63]

---

from a re-thinking of the four "position circles", encompassed by the twelve unequal hour circles (renamed "position circles") and the introduction of "two theoretical circles".

58 Santucci, "Trattato di diversi istrumenti", cit. (n. 10), fol. 4v.

59 Meucci, op. cit. (n. 2), p. 9; Righini Bonelli, "The Armillary Sphere of Antonio Santucci delle Pomarance", *Imago Mundi* 8 (1951), pp. 99–100: 99.

60 Meucci, op. cit. (n. 2), p. 8; Righini Bonelli, "The Armillary Sphere of Antonio Santucci", cit. (n. 59), p. 100.

61 Miniati, op. cit. (n. 23), p. 281.

62 *Ibid.*, pp. 281–283, identifies at least twelve craftsmen.

63 Meucci, op. cit. (n. 2), pp. 9–10; Righini Bonelli, "La sfera armillare del Santucci delle Pomarance al Museo di Storia della Scienza di Firenze", *Rivista tecnica della Svizzera*

The many parts were gathered in Santucci's temporary workshop at the Pitti Palace.[64] They were then assembled in the Maps Room of the Uffizi Gallery, where the armillary was completed on 28 May 1593. It remained there—near Egnazio Danti's terrestrial globe of 1567, moved from the Palazzo Vecchio—until at least 1769.[65]

Probably in 1771, the instrument was taken to the west wing of the second floor of the Pitti Palace, to a room where the Accademia del Cimento had held meetings a century earlier.[66] The armillary stayed there until 1775, when, by order of Grand Duke Peter Leopold of Hapsburg-Lorraine, it was moved to the new Imperial and Royal Museum of Physics and Natural History in the Torrigiani Palace.[67] In 1819, it was visible in a portico to the right of the entrance.[68] Years later, however, it was thought that the instrument would fit in a niche of the portico (which today hosts a statue of Evangelista Torricelli) opposite the entrance.[69] Both locations were in the open air, showing disregard for the item, which was employed merely for decoration.

The armillary was in very poor condition by the mid-nineteenth century. Vincenzo Antinori (1792–1865), Director of the Museum from 1829 to 1859, remarked that a thorough restoration was vital. This took place in 1876, under the direction of Filippo Parlatore (1816–1877) and the supervision of Ferdinando Meucci (1823–1893). Chemicals were used to remove the patina that was hiding graduations and painted details. The instrument was repainted locally, regilded with gold leaf and the broken rings "re-nested".[70] Finally, it was housed on the first floor of the Torrigiani Palace, in "Santucci's Room" in the Museum of Antique Instruments (*Fig. 9.6*), to the left of Galileo's Tribune.[71]

---

italiana 42, 12 (1955), pp. 292–293: 292.

64 Meucci, op. cit. (n. 2), p. 9.

65 Giuseppe Bianchi, "Catalogo dimostrativo", 1768: Florence, Archivio Biblioteca Uffizi, ms. 67, cc. 40 and 127; Pietro Mancini, "Inventario generale di tutte le preziose antichità et insigni memorie che si conservano nella Galleria di S.A.R.", 14 February 1769: Florence, Archivio Soprintendenza Beni Artistici e Storici, MS. 114, no. 79, cc. 132v–133r, no. 780.

66 Scientific instruments were in the Uffizi until 1771, when they were moved to the Royal Cabinet of Physics; see Giuseppe Bencivenni Pelli, *Saggio istorico della Real Galleria di Firenze*, Florence, per Gaetano Cambiagi Stamperia Granducale, 1779, pp. 201–202.

67 Meucci, op. cit. (n. 2), p. 3. Simone Contardi, *La Casa di Salomone a Firenze: L'Imperiale e Reale Museo di Fisica e Storia Naturale (1775–1801)*, Florence, Olschki, 2002, p. 70, n. 186, argues that the armillary was transferred directly from the Uffizi to the new Museum.

68 Anon., *Descrizione dell'Imp. e R. Museo di Fisica e Storia Naturale di Firenze*, Florence, Giuseppe Molini e Comp., 1819, p. 10.

69 Meucci, op. cit. (n. 2), p. 3.

70 *Ibid.*, pp. 7–8.

71 *Ibid.*, p. 4. The floor of the Museum of Antique Instruments was painted between 1876 and 1929, but the armillary was not moved. In 2008, the "footprint" of the instrument's

FIGURE 9.6 Cosmological model, by Antonio Santucci, Florence, 1588–1593, in the first room of the Museum of Antique Instruments at Torrigiani Palace, Florence, *c.*1876; from Ferdinando Meucci, *La Sfera Armillare di Tolomeo costruita da Antonio Santucci*, Florence, 1876.

There is no information on the move to its present location. In 1929, it was still in "Santucci's Room",[72] and was not among the first instruments forming the nucleus of the Institute and Museum of the History of Science, placed in the Castellani Palace in 1930. It was, however, in the centre of the Cosmography Room of the Museum in 1948.[73] This move caused more damage: several rings were incorrectly positioned due to the misunderstanding of their function.[74] It was also questionable whether the old repairs, made by reinforcing broken components with brass plates and screws, helped to stabilize the armillary. Marks from missing rings were also present. Many fragments were packed separately in a way suggesting that large pieces had been broken down to small ones to store them more "properly". Other fragments were lost.

Recent Restoration of the Armillary

In the early 2000s, Santucci's masterwork became the subject of close studies to ascertain its health. Its survival was mostly due to its intrinsic solidity. Following the latest thinking of his time, Santucci wanted each ring to be made of up to six beech-wood layers. The opposing tensions of the layers helped prevent distortions. X-ray images taken by the Opificio delle Pietre Dure of Florence showed that the layers were held tight by hundreds of nails (*Fig. 9.7*).

One unknown was the extent of parasitic attack. A six-month study with an ultraviolet-trap by the Institute for Valorising Wood and Arboreal Species (IVALSA) of Florence confirmed that no woodworm were alive. Ultrasound probing of several rings revealed that the wood loss due to woodworm and mould was about 10% for the meridian ring and up to about 60% for other rings.

Another unknown was the deformation of the load-bearing rings. IVALSA performed non-destructive flexion tests on ring fragments and on a five-layer beam reconstructed by the University of Florence. Finally, a mechanical arm mapped the shape of the load-bearing rings. In the case of the meridian (mean radius 1.05 m), this showed deviations of up to 6.5 cm from circular. Statistical

base was rediscovered below the cupboards of the ornithological collection stored at the Museum of Natural History of Florence. Today the footprint is hidden by carpeting.

72 Giuseppe Boffito, *Gli strumenti della scienza e la scienza degli strumenti: con l'illustrazione della Tribuna di Galileo*, Florence, Seeber, 1929, p. 209 and pl. 126.

73 Righini Bonelli, "Il Palazzo Castellani sede del Museo di Storia della Scienza", *Firenze e il mondo: rivista bimestrale di arti, letteratura e turismo* 1, 3–4 (1948), pp. 70–72: 70.

74 Righini Bonelli (ed.), *Catalogo degli Strumenti del Museo di Storia della Scienza*, Florence, Olschki, 1954, p. 48.

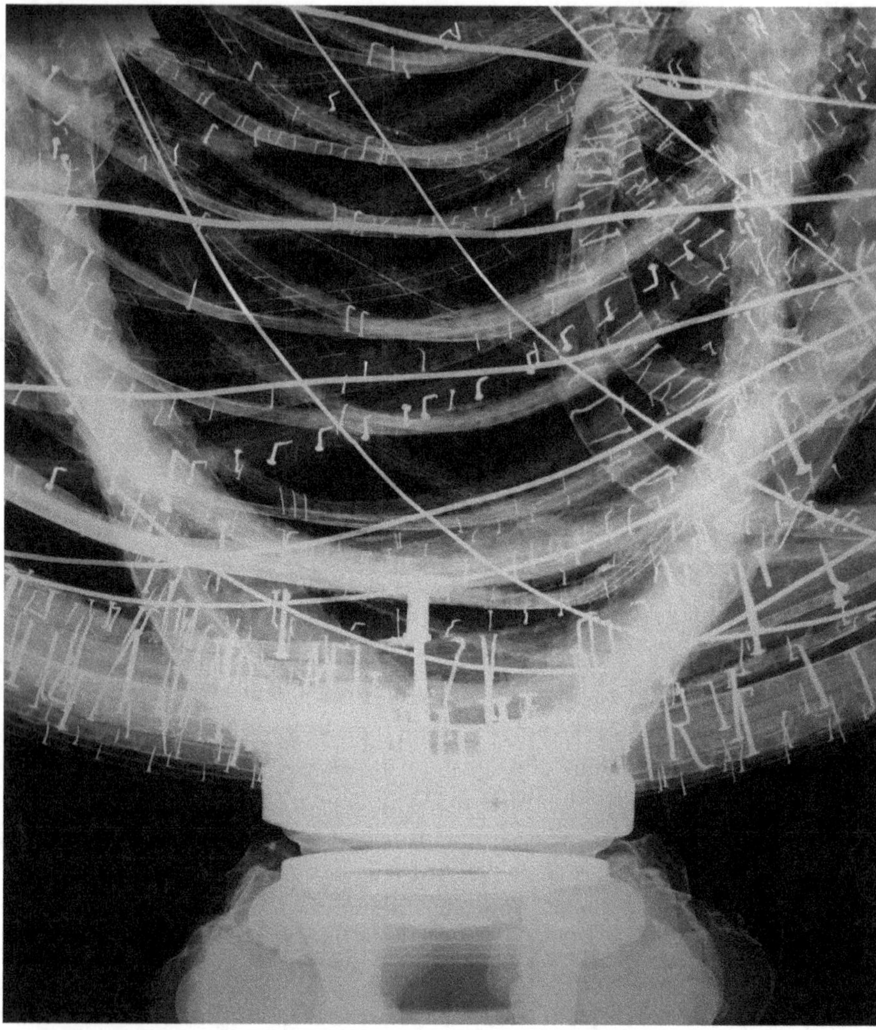

FIGURE 9.7 Cosmological model, by Antonio Santucci, Florence, 1588–1593; X-ray image revealing the wood layers of the meridian ring and hundreds of long iron nails (Museo Galileo, Florence, inv. no. 714; Photo: Opificio delle Pietre Dure, Florence).

projections revealed that, if not corrected, distortions would increase over the next 50 years and destabilize the instrument.[75]

---

75 Pierre Cassagne, *Apport de la simulation numérique pour le diagnostic de conservation des object d'art: Application à la sphère armillaire d'Antonio Santucci du 16ème siècle*, Stage de Recherche DEA, 2003, pp. 57–60.

FIGURE 9.8 Cosmological model, by Antonio Santucci, Florence, 1588–1593; restorer Roberto Buda and the central part of the armillary on a temporary stand, 25 February 2008 (Museo Galileo, Florence, inv. no. 714; Photo: Museo Galileo, Photographic Archive and Laboratory).

The new restoration took place between 2008 and 2010, during the renovation of the Museum of the History of Science. The scientific investigations had outlined a basic procedure for disassembly. The restorer Roberto Buda and his assistant Zeljka Caspar, of RelArt, Florence, consolidated parts at risk of detachment. Aided by the late Carmine Pepe of the art-transport company Arteria, they removed the top sphere with the cross and hour dial, lifted the top half-sphere cover, extracted the internal part of the instrument, and removed the bottom half-sphere and sculptured base. The last two steps revealed the wooden pin onto which the sphere could rotate around the vertical axis. About 8 cm high and 20 cm in diameter, it is remarkably small and testifies to Santucci's ability to create a well-balanced structure.

In order to detach the unequal hours' rings, the meridian was placed on a temporary stand (*Fig. 9.8*), then it was turned horizontally on four other stands. The internal components were removed by repeating a back-and-forth movement of the polar axis. Once the circles for each planetary sphere were un-nailed, shifting forth the axis freed the south pole of the sphere and then, shifting the axis back, its north pole. However, doing this was blocked by the peculiar shape of the axis: the top half is cylindrical, the bottom half pyramidal.

This cast new light on Santucci's search for a skilled blacksmith. A certain Piero di Giulio Casini shaped the axis perfectly.[76] The planetary spheres were originally attached directly to the axis, their position depending upon the size of a square hole at the south pole of each sphere. As the instrument was repeatedly moved over time, these holes widened and the planetary spheres descended. During the 1876 restoration or before, therefore, spacer cylinders had been placed between the south poles of the spheres to maintain their distances.

With the original axis temporarily replaced with a cylindrical bar, disassembly proceeded up to the sphere of Mars. The nails of the four internal spheres were still original. While the rest of the armillary was transported in single pieces, its internal part was moved en bloc to the RelArt laboratory.

The restoration aims, agreed by the restorer and the artistic and scientific supervisors (Maria Matilde Simari for the Special Superintendence to the Historic, Artistic and Ethno-Anthropologic Heritage of Florence, and the present author for the Museo Galileo), were: 1) consolidate the instrument and remove old patches, unless this endangered the armillary; 2) clean the instrument; 3) re-introduce broken parts, unless this depended on removing functional replacements; 4) restore painted elements only where their absence would be evident. The goals did not include restoration to working order, since reshaping the rings into perfect circles would have stressed the old beechwood and increased the instrument's fragility.

The armillary did not require consolidation. Santucci had wanted it made exceedingly well, something that allowed it to survive the many moves and long stay in the open air. Moreover, the woodworm holes made the instrument fragile, but relatively light. Any attempt to fill the holes with resin would have increased its weight and made it more unstable. Therefore, most of the consolidation sought to eliminate the artefacts of previous restorations. The plates and screws used to fix breakages were removed, including tin cross-plates and screws at the south poles of each planetary sphere. The plates were replaced by carefully regluing the edges of broken rings or, in a few cases, rebuilding very damaged ring sections.

Much of the restoration consisted of cleaning dust and layers of darkened shellac, which had made the original painting, gilded motifs and inscriptions illegible. Cleaning revealed colours and details throughout. Cartographic details that emerged from the terrestrial globe warrant further study. For instance, the

76 Miniati, op. cit. (n. 23), p. 282.

different hue of some coastal areas might emphasize geographic discoveries or uncertainties of longitude. Other details emerged on the zodiac band of the armillary. It became apparent that Enea Santucci was skilled at depicting human figures, but had a naive approach to animals. Inscriptions about the astrological character of the constellations and their connection with the planets were also recovered.

The cleaning was not over-done. It would have been wasteful to clean or remove nineteenth-century additions other than aged shellac. A rival restorer, for example, advised scratching away the gold-leaf applied in 1876, under Meucci's supervision, on the external rings and base, to bring the armillary to its "original" sixteenth-century "splendour". This advice was disregarded since the remains of such splendour were probably patchy and the nineteenth-century leaf-gold layer testified to a historic and homogeneous addition.

The broken rings held in store were also evaluated for reintegration. These were parts of the tropics of the sphere of the fixed stars and of the unequal hours rings. As the latter had already been replaced, the corresponding fragments were cleaned and returned to store. The tropics of the eighth sphere, which were still missing, were reintegrated. Since they were only partially preserved, missing sections were recreated with resin. This gave more sense to the instrument and was easily detectable on close inspection, but did not disrupt the perception of the whole.

As soon as the instrument was consolidated and cleaned, two other problems emerged. First, solvents had been used to remove dirt during the 1876 restoration. These had continued to act, causing partial loss of paint on the inside of the zodiac band and both sides of the ecliptic rings of the planetary spheres. The decision was made not to interfere with the ecliptic rings of the planets. Since the body of the instrument prevents a clear view of these details, visitors will not realise that the paint is partly missing. For the zodiac band, pictorial integration was considered essential due to the level of paint loss. Chromatic selection was used to distinguish original painting from new additions.

Second, most of the rings of the armillary have peculiar graduated scales, with the degrees marked by wooden teeth projecting from each side. The teeth were heavily damaged and most had lost their gold layer. It was decided to restore only completely missing teeth. For the others, no gilt repair was made, but the colour of the gilt preparation, made of white plaster, was changed to a wood colour. In this way, missing gilt is discernible by looking carefully at individual parts, but not at the whole.

In general, problem-solving followed a principle of minimal action, to guarantee the consolidation and preservation of the instrument without altering its status too much.

The Florentine Armillary and Santucci's "Trattato"

A comparison between Santucci's "Trattato" and the Florentine armillary shows that the original design for a theoretical, all-inclusive instrument changed.

Observation: The armillary has no observing function. Like the Madrid armillary, there is no magnetic compass on the base. In addition, the axis of the instrument is a solid iron bar, which cannot work as a sighting tube. The meridian ring fits in a notch in the horizon, which prevents the taking of azimuths.

Timekeeping: Since it cannot be used for observation, the armillary cannot tell the time. It can, however, work as an hour converter. All the necessary time circles are present, including those for unequal hours. The horary dial, needed for conversions, is also present, but with only one of the three hands Santucci mentions. Unless the two other hands have been lost, the remaining one indicates the astronomical time from the rotation of the internal spheres. Horary conversions would, therefore, need different procedures from those described in the "Trattato".

Astronomical computation: In principle, all the calculations explained in the "Trattato" can be performed, except those involving the stars. There is no evidence that the armillary ever had star pointers. The recent cleaning showed that the upper face of the horizon ring contains a 32-direction wind rose and solar calendar. These are barely accessible, since they are over 1.8 m above floor level and are partly covered by the balls holding the upper half of the external sphere.

All the circles needed for calculations are present. At the centre of the zodiac band, the grooved ecliptic can hold a lute string for the planetary pointers. Archival documents reveal that at least four pointers—two Suns and two Moons—were made.[77] However, the zodiac of the Florentine armillary contains neither a second groove for the Moon's orbit, nor a large internal ring for determining the dates of new and full Moon, both found in the Madrid instrument.

Besides its close agreement with the "Trattato", it is inconceivable that the massive Florentine armillary could have been rotated around the vertical and in the meridian plane to perform calculations. Rotation in the meridian plane is impossible due to the mount of the altitude circle (fixed for the latitude of Florence), and the peculiar structure of the central terrestrial globe. X-ray images revealed that the globe contains an eccentric mass. When the sphere is rotated with a crank, the mass acts as a counterweight, keeping Florence at the top of the terrestrial globe. If the latitude is changed, the mass would

77 *Ibid.*, p. 283.

destabilize the instrument. From a practical point of view, therefore, the armillary can only perform calculations for the latitude of Florence and depending on the simple daily rotation of the sphere.

Education: The most significant difference between the "Trattato", the Madrid armillary and the Florentine armillary concerns the seven sets of rings nested in the space between the terrestrial globe and the sphere of the stars. Each set, made of six rings (equator, ecliptic, colures and polar circles), models a planetary sphere. With this addition, the Florentine armillary demonstrates the structure of the cosmos, rather than allowing the user to identify the planets and brightest stars, or perform calculations for the dates of full and new Moon, although the representation is rudimentary, closer to the Platonic than to the Aristotelian or Ptolemaic concept. In contrast to the introductory chapter of the "Trattato", the structure is over-simplified. The ninth sphere of the prime mover and the crystalline sky are both missing. In addition, the spheres of the planets can only move together, at the same rate, around the polar axis. The double pointers for the Sun and Moon suggest that another set of pointers could be placed around the ecliptic rings of the planetary spheres in addition to a set for the zodiac band. The educational purpose of the instrument extends to theological aspects of the cosmos. The outermost sphere models the Empyrean sky, the domain of the Holy Father (depicted inside the top of the sphere) and the angelic ranks (four sculptured angels hold the horizon ring).

Astrology: The Florentine armillary is a more complete astrological compendium than the one in Madrid. All the information for the zodiacal signs given in the "Trattato" is found on the instrument. For example, Santucci writes that Sagittarius is "masculine [...] fiery, hot and dry".[78] The restoration has revealed that the corresponding 30° portion of the zodiac of the armillary has a bright red ground colour for the element of fire. It also includes the label "Sagittarius: fiery, masculine, hot and dry sign". The instrument goes beyond the "Trattato", however: following Ptolemy's *Tetrabiblos*, it includes a label identifying Sagittarius as the "diurnal house of Jupiter" (*Fig. 9.9*). The influences of the planets when passing through the signs are noted around the whole zodiac.

The armillary can perform all the astrological functions described in the "Trattato". In particular, in order to draw a nativity chart, the rings of the houses are distinguished from those which complete the sequence for unequal hours. In contrast to the "Trattato" and the Madrid armillary, the distinction does not rely upon the shape and graduation, but on the colour: the rings of the houses are gold and red, the others gold and black. Again, unlike the Madrid armillary but copying the "Trattato", the Florentine armillary has an ad hoc additional

78 Santucci, "Trattato sopra la nuova inventione", cit. (n. 20), fol. 9v.

FIGURE 9.9 Cosmological model, by Antonio Santucci, Florence, 1588–1593; detail of Sagittarius showing astrological information (Museo Galileo, Florence, inv. no. 714; Photo: Museo Galileo, Photographic Archive and Laboratory).

device to indicate which planets rule the unequal hours. A ring divided into 24 equal segments associated with the planets can slide above the sphere's celestial equator. As 24 is not divisible by 7, the descending sequence of the planets is repeated three times, with the remaining three places allocated to Venus, Mercury and the Moon. It is likely that the users were supposed to look only at the nocturnal part of the sliding ring, with the remainder invisible, over 1.8 m above the floor.

With increasing experience in mathematics, Santucci would have realized other limits of the original design. The "Trattato" includes four position circles with poles at the intersections between the horizon and the local meridian. The circles are supposed to divide the celestial equator into twelve equal parts, having a close relationship with the celestial houses.[79] This geometrical mistake led Santucci to discard the four rings, or rather to replace them with the six rings of the celestial houses determined according to the method ascribed to Campanus of Novara, based on the division of the prime vertical into twelve equal parts. Departing from the "Trattato", the Florentine armillary has the altitude circle revolving in the space between the sphere of the fixed stars and

79 *Ibid.*, fol. 16r.

the rings of the unequal hours. The ring has fixed pins on the meridian, at the zenith and nadir of Florence, and carries a second ring revolving around a horizontal axis. If the user places the altitude ring in the plane of the meridian, the second ring can be positioned on any degree of the celestial equator. It can, therefore, mark the beginning of the twelve houses by Regiomontanus's method, based on the division of the celestial equator into twelve equal parts, as originally planned for the four discarded rings.[80]

Conclusion

It seems that an involution-evolution process took place from the "Trattato", to the Madrid armillary, to the Florentine armillary. In the last instrument, provided with all the funds needed, Santucci did his best to accommodate the Grand Duke's wishes. Ferdinand had the ambitions not of an expert astronomer, but of a learned man who wanted an astonishing model of the cosmos to display in his Uffizi Gallery. The instrument had to fit into the new context of the Maps Room.[81]

The model had to be impressive, beautiful, evocative of the cosmos as a whole, with its many spheres, and revolve—using a crank inserted at the bottom of the polar axis—to amaze important visitors to the Gallery. It also needed to evoke Medici knowledge and power. Within this context, observation, timekeeping, hour conversions and calculations were marginal functions. Astronomical and astrological inscriptions were decorative rather than instructive. Even the conceptual flaws were destined to disappear in the overwhelming artistic framework.

As the building of the large armillary progressed, the instrument gained two other functions not found in either the "Trattato" or the Madrid armillary. The Florentine instrument is surmounted by a cross, the symbol of Christianity, associated both with Ferdinand's former position as a cardinal and, as Grand Duke, defender of the Holy Roman Church. In addition, following long negotiations, Ferdinand married Christina (1565–1636), the daughter of Charles III, Duke of Lorraine, on 25 February 1589. The armillary celebrated the union: the

---

80 North, op. cit. (n. 41), pp. 27–30.

81 See Filippo Camerota, "The Medici Collection of Mathematical Instruments: History and Museography", in Giorgio Strano, Stephen Johnston, Mara Miniati, Alison Morrison-Low (eds.), *European Collections of Scientific Instruments, 1150–1750*, Leiden, Brill, 2009, pp. 129–148: 136–138.

FIGURE 9.10 Cosmological model, by Antonio Santucci, Florence, 1588–1593; detail of the coats of arms of the Medici and Lorraine families (Museo Galileo, Florence, inv. no. 714; Photo: Museo Galileo, Photographic Archive and Laboratory).

polar plaques of the unequal hours rings bear the coats of arms of the Medici and Lorraine families beneath the Grand Ducal crown (*Fig. 9.10*). Below the cross on top, exactly at the zenith, another plaque represents the Lord blessing all that lies below: the celestial and terrestrial worlds and union of the two Ducal houses. Within the sphere, cosmology, astronomy, astrology and geography form a compound of mathematical disciplines at the service of the Medici crown.

The finished structure of the Florentine armillary was very different from the description in the "Trattato". It was no longer an "armillary sphere", but a "sphere of marvellous size and the most rich ever seen, of which, if our Lord will concede it to me, a large compendium will soon appear, explaining all its parts and all the astronomical and geographical operations it performs".[82] In 1593, as he began the "Trattato di diversi istrumenti matematici", it seems that Santucci was planning a thorough revision and printing of the 1582 "Trattato", to accommodate the new features of the grand cosmological model. Regrettably, the revised treatise never came to fruition.

---

82 "[...] una sfera di maravigliosa grandezza e la più copiosa che si vedessi già mai della quale se il Signore i Dio celo concederà se ne vederà presto un copioso compendio, dove si dischiarano li sua termini con tutte le operationi astronomiche, et geografiche che in quella si contengano"; Santucci, "Trattato di diversi istrumenti", cit. (n. 10), fol. 4v.

CHAPTER 10

Were Globes Used in the Practice of Early Modern Astrology?

Jim Bennett and Sylvia Sumira

The purpose of this chapter is to look for evidence for the popular association between astrologers and celestial globes, focusing on the sixteenth and seventeenth centuries.[1] Early printed globes were not only decorative furnishings or educational accessories, but were instruments used for calculation. This is true of both terrestrial and celestial globes and many books explaining their use were published from the sixteenth to the eighteenth century. In the case of the celestial globe, such applications were popularly associated with astrology. A number of pictures from the period, which contain celestial globes, have been given titles suggesting that they depict astrologers. It can be difficult to be certain that astrology was intended, not least because it is problematic in a portrait to differentiate astrology from astronomy, a distinction that, in any case, is not a sharp one in the sixteenth century and to some extent the seventeenth as well.[2]

What evidence have we for thinking that the "ordinary" celestial globe was, at least in part, an astrological instrument? By "ordinary" we mean to exclude globes with additional printed astrological information, beyond the stars in

1 For a rich account of the background, see Elly Dekker, "Globes in the Service of Astrology", in *Illustrating the Phaenomena: Celestial Cartography in Antiquity and the Middle Ages*, Oxford, Oxford University Press, 2013, pp. 388–408.

2 Since presenting a paper at the XXXIV Scientific Instrument Symposium in Turin in September 2015, we have been given relevant information and references by a number of delegates. Those that fall within our specific brief of early-modern printed celestial globes are referred to in the body of the article. We are grateful in addition to Josefina Rodríguez-Arribas for information on references by Alfonso X, King of Castile and Leon (1221–1284) on the use of globes in astrology, see Alfonso X of Castile, *Libros del saber de astronomía*, ed. by Manuel Rico y Sinobas, Madrid, 1863–1867, 5 vols.; v. 1, pp. 206–208. We are grateful also to Samuel Gessner for information on the astronomical clock by Philipp Immser, 1555–1557, in the Technisches Museum für Industrie und Gewerbe, Vienna, inv. no. 11939, where a system for delineating the houses is accommodated to a rotating globe; see Paolo Galluzzi (ed.), *Galileo: Images of the Universe from Antiquity to the Telescope*, Florence, Giunti, 2009, p. 322; Henry C. King, *Geared to the Stars: The Evolution of Planetariums, Orreries and Astronomical Clocks*, Bristol, Adam Hilger, 1978, pp. 68–72.

 | DOI:10.1163/9789004381438_012

their constellations, the circles for the equator, ecliptic and tropics, and the zodiacal calendar on the horizon ring. Such information could, of course, be used in either astronomy or astrology. But a few globes are more explicit. Elly Dekker, for example, points out that Gemma Frisius includes planetary symbols of stars, indicating their associated planetary natures, on his celestial globe.[3] She suggests that it may have been Gemma's professional interest in medicine that led him into globe-making for the medical application of astrology.[4] She points also to the astrological information given on the horizon ring of Mercator's celestial globe.[5] We take it that content such as this is a clear indication of an astrological intent, even if the extent of its use is a different question.[6]

Globes seem to fit most naturally into the tradition of cosmography, and the same is true of their early makers. Cosmography was the description and representation of the cosmos as a whole, with a particular emphasis on the relationship between the Earth and the heavens. This means that the matched pair of globes, terrestrial and celestial, first produced by Johann Schöner in 1515, was particularly well accommodated to the discipline. The cosmographical context for the globe is indicated also by the production, first by Gemma, followed by Mercator, of a terrestrial globe with astronomical content, notably the ecliptic and a number of stars, which Dekker calls a "cosmographic globe".[7] The fact that most globes (even the "cosmographic" globes of Gemma and Mercator) were originally one of a pair, a tradition maintained through the eighteenth century (and in decline in the nineteenth), also suggests a cosmographical identity. While a concern with the relationship between Earth and heaven would be consistent with the basic tenets of astrology, all the technical content of astrological practice falls outside cosmography and within a specialism of its own. Taken together, these observations tend not to place the primary disciplinary identity of the globe in astrology.

---

3 For the Gemma Frisius celestial globe, see Elly Dekker, *Globes at Greenwich. A Catalogue of the Globes and Armillary Spheres in the National Maritime Museum, Greenwich*, Oxford, Oxford University Press/National Maritime Museum, 1999, pp. 87–91 and 341–342.

4 Elly Dekker, "Globes in Renaissance Europe", in David Woodward (ed.), *The History of Cartography*, vol. 3, pt. 1: *Cartography in the European Renaissance*, Chicago, University of Chicago Press, 2007, pp. 135–173: 153–155 (astrology) and 154 (Gemma Frisius and medicine).

5 Dekker, *op. cit.* (n. 3), p. 415. See also Annelies van Gijsen, "De astrologie", in Marcel Watelet (ed.), *Gerardus Mercator Rupelmondanus*, Antwerp, Mercatorfonds, 1994, pp. 220–233; Steven Vanden Broecke, "Dee, Mercator, and Louvain Instrument Making: an Undescribed Astrological Disc by Gerard Mercator (1551)", *Annals of Science* 58 (2001), pp. 219–240.

6 Also useful is E. Dekker, *Istituto e Museo di Storia della Scienza. Catalogue of Orbs, Spheres and Globes*, Florence, Giunti, 2004, pp. 172–175.

7 Dekker, op. cit. (n. 4), pp. 141–148.

Nonetheless objects can be relevant to different disciplines, so what of the pictorial images of "astrologers" with single, unpaired, celestial globes? If the overlap with astronomy makes an unequivocal identification difficult, we might look for other professionals depicted with globes, where they can have used them only for astrological purposes, even though that was not their primary vocation. We find celestial globes in paintings of alchemists and physicians, who are usually shown in the course of their work. If a globe was used in such a context, this must have been for an astrological procedure, such as finding the propitious time for an alchemical operation, or prognosticating the course of a fever and determining the appropriate time for a therapy. Or might a globe be present simply to indicate learning?[8]

Documentary Evidence

Occasionally we can find an unambiguous example. A woodcut from Robert Fludd's book *Utriusque Cosmi Historia* of 1617 shows an astrological consultation (*Fig. 11.4* in the next chapter). There is no doubt that we see an astrologer casting a horoscope for his client, and among the very few tools on the table is a celestial globe. It might still be said that an image is one thing, but how can we be sure that this really happened, and with any regularity? For that we would need documented instances and, in the nature of astrological practice, they may not be easy to find.

One such instance is recorded in a letter written by the famous English astrologer William Lilly to Elias Ashmole in December 1664, on the subject of a comet he has observed:

> I first had sight of it ☽ 26. Dec. 6h 45′ P.M. I set my figure by the Globe—and had 2: degr: of ♉ Culminating—and about 17. ♌ in the Ascend[ant]—if you observe under Caput Medusae—you will see it:—it is about 14. of ♉ and hath about 16. deg. north Lat: […].[9]

8 Such images are not difficult to find. For example there are a number on the website of Art UK, http://artuk.org/ (accessed 31 March 2018); for example, nos. 125755, 125756, 125757, 125758, 125969. See also the chapter by Richard Dunn in this book, *infra*, pp. 230–252.

9 Conrad Hermann Josten, *Elias Ashmole (1617–1692): His Autobiographical and Historical Notes, his Correspondence, and other Contemporary Sources Relating to his Life and Work*, Oxford, Oxford University Press, 1966, 5 vols.; v. 3, p. 995. For Lilly, see Ann Geneva, *Astrology and the 17th-Century Mind: William Lilly and the Language of the Stars*, Manchester, Manchester University Press, 1995. See also Patrick Curry, *Prophecy and Power: Astrology in Early Modern England*, Cambridge, Polity Press, 1989, pp. 35–57.

The least that Lilly must have done was to set his globe for his latitude and the date and time he gives. He would first adjust the altitude of the pole (equivalent to the latitude) by turning the meridian ring, so that the given latitude value on the degree scale on the north side of this ring coincides with the north point on the horizon ring. From a zodiacal calendar (often printed on the paper surface of the horizon ring) he could find the position of the Sun in the zodiac for the date, and could then locate that point on the ecliptic circle on the globe, and rotate the globe to bring that zodiacal position into the plane of the meridian (the easterly face of the meridian ring). He would then bring to the same plane the friction-tight index that sits above the hour circle at the pole, so that it indicates 12 o'clock. This index is fitted to the polar axis, so rotates with the globe. When it is set to the given time, the globe will show the sky as it was when the comet was sighted, and Lilly finds Leo 17 on the eastern horizon and 2 degrees of Taurus on the meridian.

Of course the *position* Lilly gives for the comet is a different matter, and simply had to be observed. Presumably he knew the sky sufficiently well to note its appearance in the head of Medusa (in the constellation of Perseus), and the globe will give him that position in ecliptic co-ordinates: Taurus 14 and 16 degrees north.

Here we have an instance of an astrologer using a celestial globe to perform a calculation. Does Lilly's letter imply that he took this further, when he says, "I set my figure by the Globe"? Having found the ascendant, has he then drawn his figure of the houses of heaven and added to it the ecliptic positions of the cusps of the twelve houses after reading them off the globe?[10] The houses divide the local sky, based on the horizon, into twelve zones by great circles passing through the two intersections of the horizon and the meridian ring (the north and south points on the horizon). The figure or horoscope is a schematic representation of these houses (*Fig. 10.1*). The limits or cusps of the houses are generally given as the points where these great circles intersect the ecliptic.

Certainly the cusps of the houses could be found with a globe. The most accessible and thorough account of how to do this, one close in date to Lilly's report, comes from Joseph Moxon in his *Tutor to Astronomie and Geographie* (first edition 1659), much of it derived from the earlier work of Willem Blaeu.[11]

10 For the houses, see John David North, *Horoscopes and History*, London, Warburg Institute, 1986, pp. 1–69.

11 Joseph Moxon, *A Tutor to Astronomie and Geographie*, London, Joseph Moxon, 1659. In what follows we have generally used the edition of 1670.

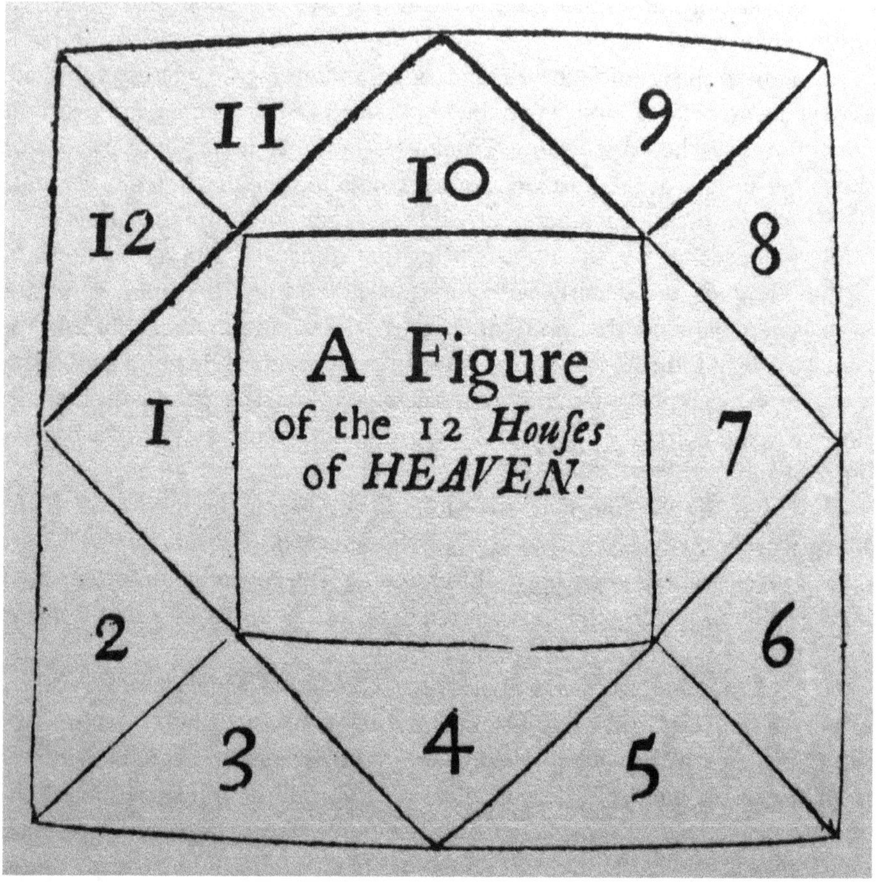

FIGURE 10.1 The figure for the houses of heaven, numbered anticlockwise from the eastern horizon; from Joseph Moxon, *A Tutor to Astronomie and Geographie*, London, Joseph Moxon, 1670.

This is an expanded version of Moxon's earlier translation of Blaeu's treatise on the use of the globes, which Moxon had published in 1654.[12]

Moxon played many parts: printer, globe maker, type-founder, instrument maker (in paper), hydrographer, writer, translator, publisher, mathematician, Fellow of the Royal Society and astrologer.[13] According to Ashmole, in 1682–

12 Willem Blaeu, *A Tutor to Astronomy and Geography*, trans. by Joseph Moxon, London, Joseph Moxon, 1654.

13 David J. Bryden, "Moxon, Joseph (1627–1691)", *Oxford Dictionary of National Biography*, Oxford University Press, 2004, www.oxforddnb.com/view/article/19466 (accessed 26 July 2015); Derek Albert Long, *'At the sign of Atlas': The Life and Work of Joseph Moxon, a Restoration Polymath*, Donnington (Lincolnshire), Shaun Tyas, 2013.

1683 Moxon briefly revived the Society of Astrologers, or at least its annual feast; Lilly had been an influential figure in its heyday, 1649–1658.[14] Moxon's engraved portrait includes two astrology books he published, *A Tutor to Astrologie* by William Eland and *Christologia* by John Butler, where horoscopes are created for the life of Christ.[15] In Moxon's own book on globes, the *Tutor* of 1670, he considerably expands the astrology content offered by Blaeu. It is characteristic of the enterprising Moxon that the Eland volume is "Digested into this pocket-volumn, for the conveniency of those that erect figures abroad".[16]

In his *Tutor*, Moxon deals with two methods of dividing the heavens into houses: according to Johannes Regiomontanus and to Campanus of Novara (Blaeu deals only with the Regiomontanus method).[17] In either, the cusps of the four "cardinal" houses are found immediately when using a globe, since they coincide with the horizon and meridian rings. The cusps of the other eight houses are found by stepping off successive sections of 30 degrees. In the Regiomontanus system, the steps progress along the equator and the positions are noted where the great circles containing these points and the north and south points on the horizon meet the ecliptic. This is done using a fitting on the globe, which Moxon sometimes calls the "semicircle of position" and sometimes the "circle of position".[18] He describes this at the beginning of the book as a brass semicircle pivoted at the north and south points of the horizon ring (*Fig. 10.2*).[19] So, without rotating the globe, the semicircle of position can be moved to the 30-degree steps on the equator and the corresponding positions read off the ecliptic. By reading the four remaining cusps above the horizon and noting their opposite points on the ecliptic, the task can be completed in four steps.

---

14 Keith Thomas, *Religion and the Decline of Magic: Studies in Popular Beliefs in Sixteenth and Seventeenth Century England*, London, Weidenfeld and Nicolson, 1971, p. 304; Bernard Capp, *Astrology and the Popular Press. English Almanacs 1500–1800*, London, Faber, 1979, p. 186; Josten, op. cit. (n. 9), v. 4, pp. 1705 and 1712.

15 William Eland, *A Tutor to Astrologie*, London, Joseph Moxon, 1670; John Butler, *Christologia, or, a Brief (but True) Account of the Certain Year, Moneth, Day and Minute of the Birth of Jesus Christ*, London, Joseph Moxon, 1671. For Moxon's portrait, see the website of the National Portrait Gallery, London, www.npg.org.uk/collections/search/portrait/mw138772/Joseph-Moxon?LinkID=mp88178&search=sas&sText=Joseph+Moxon&role=sit&rNo=0 (accessed 30 March 2018).

16 Eland, op. cit. (n. 15), title page.

17 J. Moxon, *A Tutor to Astronomie and Geographie*, London, Joseph Moxon, 1670, pp. 122–135.

18 Both terms are used, somewhat indiscriminately, in the primary literature; we will adopt "circle of position" for the geometrical concept of part of a great circle, and "semicircle of position" for a physical arc of brass.

19 *Ibid.*, p. 7.

FIGURE 10.2 The circle of position, "CIRCVLVS POSICIONIS", on a celestial globe; from Johann Schöner, *Globi stelliferi, sive sphaerae stellarum fixarum usus*, Nuremberg, Joannes Petreius, 1533.

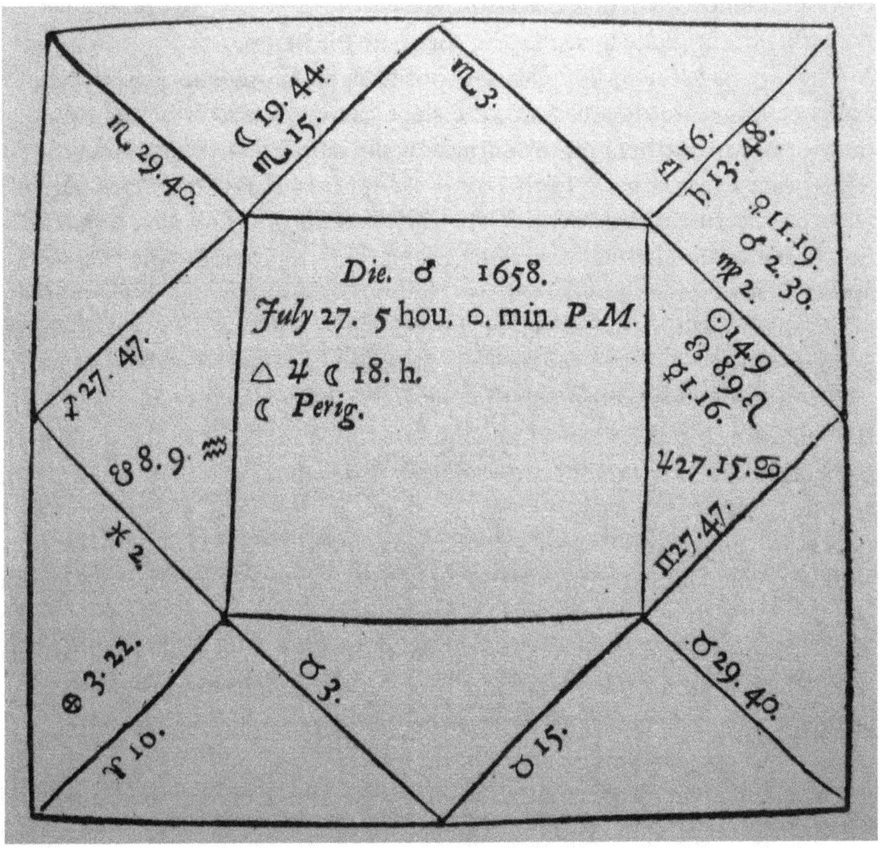

FIGURE 10.3 The figure with the addition of the ecliptic longitudes of the cusps of the houses, the planets, the nodes of the Moon and the Part of Fortune; from Joseph Moxon, *A Tutor to Astronomie and Geographie*, London, Joseph Moxon, 1670.

For finding the boundaries of the houses in the system of Campanus, the globe from the horizon to the meridian is divided into three equal segments. A further accessory, a brass quadrant arc called the "quadrant of altitude" is clamped to the meridian ring at the zenith. The semicircle of position is then set to successive steps of 30 degrees on the quadrant and the cusps of the houses read from the ecliptic circle on the globe.

The zodiacal positions of the cusps are added to the boundaries of the houses on the figure, which must then be completed by adding the planets and the lunar nodes to the houses where they are found at the time in question (*Fig. 10.3*). But there are no planets on a globe. As Moxon says, "Planets and Comets cannot be placed on the Globe so as their places will long retain correspondence with their places in Heaven".[20] Yet he is keen to demonstrate the utility of the globe by using it to complete the figure.

Once we have found the planetary positions from ephemerides, in equatorial or ecliptic co-ordinates, since the globe can operate in either, Moxon says that we can mark them on the surface of the globe, and these marks can be used at least for a time.[21] In the cases of Jupiter and Saturn, this could be a week or a fortnight, he says, though much less for other planets. Some positions have to be altered "very often," especially the Moon, which moves by about a degree of longitude every two hours. Blaeu also explains the marking of planetary positions on the globe, but for him it seems only for the moment in question—he does not say they can be useful for a period of time.[22]

When Moxon works through the detailed steps needed to "erect a figure of Heaven" using a globe, his recommendation is that marking on the surface the planets and the lunar nodes comes first.[23] This is done in either equatorial or ecliptic co-ordinates, using either the equator on the globe and the meridian ring, or the ecliptic circle set on the horizon ring and the quadrant of altitude with its pivot clamped at the ecliptic pole. The planetary positions should not be marked directly from those given in the ephemerides, which are for noon (they are adjusted, of course, for local longitude and local time), but the planet should be placed for the actual time (not the noon equivalent time), this adjustment being calculated as a proportion of the difference between the positions at successive noons given in the tables.[24]

The globe is then set for latitude, date or solar declination, and time, and fixed, Moxon says, with a quill, presumably used as a wedge between the globe and the horizon ring. It was fixed because, after he has noted the ascendant or point of the ecliptic on the eastern horizon, and recorded the opposite ecliptic position, which is the cusp of the seventh house, he will use his "Circle of Position" to relate the 30-degree equatorial steps to the corresponding degrees of the ecliptic. Because the houses are numbered anti-clockwise, the first such position to be found is the cusp of the twelfth house, and its opposite is that of the sixth, then the eleventh and fifth. The cusp of the tenth house is found from the meridian ring, and Moxon says that thereafter, proceeding towards the western horizon, involves "taking the Semi-Circle of Position off its Poles,

20 *Ibid.*, p. 105.

21 *Ibid.*, pp. 105–106.

22 W. Blaeu, *Institutio astronomica de usu globorum & sphærarum cælestium ac terrestrium*, Amsterdam, Johan Blaeu, 1668, pp. 109–112.

23 Moxon, op. cit. (n. 17), p. 123.

24 *Ibid.*, p. 126.

I place it on the West side of the Meridian".[25] That gives us a clue to how the semicircle of position was mounted. If it is outside the meridian ring, it cannot clear the hour circle; if within, it will be obstructed by the polar axis. Either way, it will have to be taken off and remounted for the houses to the west of the meridian ring.

The ecliptic positions in longitude of the cusps are added to the figure. Curiously, despite the great detail Moxon has given in going through the procedure thus far, he does not mention the final step of reading the longitudes of the planets and lunar nodes and adding these to the figure, as they are placed in the appropriate houses. Did he simply forget to mention this step? Or is this a clue that, in truth, planetary positions would not be found using a globe?

Moxon discusses several further astrological applications of the globe. Among these, he first deals with the length of the planetary hours for latitude and date—planetary, or unequal, hours being used, he says, by astrologers—and then with finding the ruling planet for each hour, day and night.[26] Next he explains locating what astrologers call the 'Part of Fortune', which involves taking the angular distance between the Sun and the Moon and adding it to the ascendant in the direction of the houses (Blaeu does not include this). It is then that he appears to remember the need to mention adding the planets to his figure, saying "the character [...] which represents the Part of Fortune, I set in its proper place of the Figure, as I did the Planets".[27]

The circle or semicircle of position might be taken as an important indicator of the astrological associations of celestial globes. As Moxon states emphatically, it was "for the finding the twelve Astrological Houses of Heaven".[28] Its use is explained in early treatises on the use of the globe. Gemma Frisius, for example, describes and illustrates it in his *De principiis Astronomiae et Cosmographiae*, first published in 1530, also saying that it was for setting out the houses of heaven.[29] The pioneer of the printed celestial globe, Johann Schöner, dealt with it in his tract of 1533 on his new celestial globe.[30] In England, we

---

25 *Ibid.*, p. 127.

26 *Ibid.*, pp. 131–132.

27 *Ibid.*, pp. 132–133. There seems to be a mistake here by the block cutter or designer, as in the Regiomontanus figure there are two Suns, while in the Campanus figure we do see the 'Part of Fortune'; *ibid.*, pp. 127 and 129. This is not the case in all editions.

28 *Ibid.*, p. 7.

29 Gemma Frisius, *De Principiis Astronomiae et Cosmographiae, deque usu globi ab eodem editi*, Paris, Gulielmum Cauellat, 1556, pp. 47–49; on the semicircle of position, see also *ibid.*, pp. 49–51 and 131.

30 Johann Schöner, *Globi stelliferi, sive sphaerae stellarum fixarum usus*, Nuremberg, Joannes Petreius, 1533; also in Johann Schöner, *Opera mathematica*, Nuremberg, Montanus & Neuber, 1561, note fols. 140 and 164v–165; John W. Hessler, *A Renaissance Globemaker's*

find Thomas Hood in 1592 listing it among the parts of a globe and explaining its use for finding the houses.[31] Thomas Blundeville, in his *Exercises*, first published in 1594, describes a pair of globes by Mercator, including a detailed account of using the semicircle of position in setting a figure and finding the houses.[32] Blundeville was a friend of John Dee and we know that Dee acquired such a pair when he visited Gemma and Mercator in Louvain around 1548.[33]

Blundeville says that the semicircle is "fastned on the Horizon, so as it may be removed and set upon which side of the Globe you will".[34] In the worked example he gives, he talks about the semicircle "being fastened in its due place of the East side of the Horizon" and later being moved to the west side.[35] He mentions that the globe is to be stopped from turning "with some prettie wedges of woode being thrust betwixt the Horizon & the body of the Globe".[36] Details such as this give some reassurance that he is describing a real operation.

Mercator himself, in his instructions for the use of his astrological disc, advises users to find the cusps of the houses from his celestial globe before transferring these to the disc:

> He should seek the beginnings of the twelve houses of the heavens on the globe for the given time and latitude of the place, and, if he wishes, he should mark these with needles on the rim of the instrument [i.e. Mercator's astrological disc].[37]

Blundeville thinks that setting a figure by the globe ends with locating the cusps of the houses on the ecliptic (using the Regiomontanus division). For him, placing the planets in their houses is not an operation for the globe: "And if you would know what planets should be placed in everie house, you

Toolbox: Johannes Schöner and the Revolution of Modern Science 1475–1550, Washington, Library of Congress, 2013, pp. 113–151.

31 Thomas Hood, *The Use of Both the Globes, Celestiall, and Terrestriall*, London, Thomas Dawson, 1592, sigs. C6 and H6-H7.

32 Thomas Blundeville, *M. Blundevile his Exercises*, London, John Windet, 1594, fols. 232–234.

33 James Orchard Halliwell et al. (eds.), *John Dee's Diary*, Cambridge Books Online, http://ebooks.cambridge.org/chapter.jsf?bid=CBO9781139245791&cid=CBO9781139245791A024 (accessed 20 November 2015).

34 Blundeville, op. cit. (n. 32), fol. 207.

35 *Ibid.*, fol. 233v.

36 *Ibid.*

37 Vanden Broecke, op. cit. (n. 5), p. 239.

must learne that out of the Ephemerides, or out of some of the Astronomicall Tables".[38]

Although the circle of position is commonly mentioned in early treatises, to date we have found only Blaeu and Moxon suggesting adding planets to the globe in manuscript. Is there really any point in that final step? We might imagine advantages of accuracy or of convenience.

On the horoscope the planets are placed in their houses and their positions in the zodiac, that is, their ecliptic longitudes, are noted. The houses are based on the horizon: they divide the *local* sky and their boundaries are not parallel to lines of ecliptic longitude. This means that a planet's displacement in latitude from the ecliptic will preserve its ecliptic longitude, but just might cause it to stray into an adjacent house, not the one indicated by its longitude on the ecliptic. Of course the planets maintain low latitudes, so this will not happen often, but there is, we might say, a "rightness", an appearance of astronomical accuracy and rigour, in being concerned about this possibility. Could there be an advantage of convenience? The direct reading made possible by the planets' positions being marked on the globe could be considered a convenience but would that offset the inconvenience of marking them there in the first place?

Material Evidence

Two lines of evidence gathering emerge from this look at early printed treatises on globes, in particular Moxon's *Tutor*. First, are there surviving globes with semicircles of position? Or are there examples where the semicircle is missing but evidence of a mounting or fitting arrangement survives? We might take an answer to this question as indicative but not conclusive with regard to astrological use. Without the semicircle, a globe could still be used to find the ascendant. In fact a user might, by turning the globe through the successive steps of 30 degrees, note the cusps of the houses, using the horizon ring as an index (at least in the Regiomontanus system). On the other hand, an owner might have a semicircle and not use it at all.

The second line of evidence gathering, which could be more conclusive but more difficult to find, would be to ask whether any globes show signs of having been marked with planets or lunar nodes, even temporarily.

We have seen that some authors who were also makers, such as Schöner, Gemma Frisius, Blaeu and Moxon, describe the use of the semicircle of position

38 Blundeville, op. cit. (n. 32), fol. 234.

for locating the houses of heaven.[39] So we might expect to find evidence of the feature on some of their globes. Of course, the semicircle was not permanently attached, but placed on the globe when needed, and had every chance of becoming lost.

A celestial globe by Mercator at the Scheepvaartmuseum, Amsterdam, survives with a semicircle of position (*Fig. 10.4*).[40] This is the 42-cm globe of 1551. An example of the same globe is in the collection of the Historisches Museum Basel, also with a semicircle of position.[41]

Since semicircles could easily become lost, do any other surviving globes have traces of once having had such accessories? We should first discover what we can about how semicircles were mounted. The most famous relevant image is of a globe by Schöner, included in the double portrait known as "The Ambassadors", painted in 1533 by Hans Holbein the Younger and now in the National Gallery, London. The upper shelf of objects includes a number of mathematical instruments, one of them the celestial globe Schöner introduced the same year. Holbein's extraordinary attention to detail allows us to see a pivot pin held by a fastening, perhaps a type of clip, embracing the meridian ring.[42] The pin is held as closely as possible to the east side of the ring, which is the plane of the meridian, and the semicircle moves within the ring (though above the thinner quadrant of altitude), but of course has to be taken off to be refitted on the western side of the meridian.

The same fitting is seen in a woodcut accompanying Schöner's account of the globe as it appeared in his *Opera Mathematica* of 1551 (*Fig. 10.5*). At first sight it appears here to be on the western side, which would take the pivot pins farther from the plane of the meridian, but the block-cutter has reversed the image left to right, even though the lettering has been cut correctly. The constellations are facing in the wrong direction.

---

39 Although not falling within the precise parameters of this study, we should note the different arrangement for finding the houses by a globe, using a set of semicircles, that occurs uniquely on the manuscript globe made by Johann Stöffler in 1493; see Dekker, op. cit. (n. 4), p. 154.

40 Scheepvaartmuseum, Amsterdam, inv. no. A.1175(01), www.maritiemdigitaal.nl/index.cfm?event=search.getdetail&id=101004336 (accessed 3 April 2018); none of the images shows the semicircle of position, but it is extant.

41 Historisches Museum Basel, inv. no. 1879.34; Sabine Söll-Tauchert, Raphael Beuing, Burkhard von Roda (eds.), *Die grosse Kunstkammer: Bürgerliche Sammler und Sammlungen in Basel*, Basel, Christoph Merian Verlag, 2011, pp. 192–193. We are grateful to Raphael Beuing of the Bayerisches Nationalmuseum for bringing this globe and reference to our attention.

42 The image online at www.nationalgallery.org.uk/paintings/hans-holbein-the-younger-the-ambassadors (accessed 3 April 2018) may be enlarged to show this feature clearly.

FIGURE 10.4 Celestial globe, by Gerard Mercator, 1551, with a detached semicircle of position (Scheepvaartmuseum, Amsterdam, A.1175(01).
PHOTO: SYLVIA SUMIRA.

There is no reason to think that this represents the only possible attachment. Blaeu mentions rotating the semicircle on a pair of poles or pinnules.[43] Moxon gives a little more detail in his own treatise, saying of the semicircle of position:

43 Blaeu, op. cit. (n. 22), p. 13.

FIGURE 10.5 The fitting for the semicircle of position; from Johann Schöner, *Opera Mathematica*, Nuremberg, Montanus & Neuber, 1551 (Library of Congress, QA33. S28 1551).

> at the two ends there is an Axis, which is fitted into the two holes of two small studs, fixed in the North and South points of the upper Plain [sic] of the Horizon: upon this Axis it is moved up and down, according to the intent of your operation.[44]

This suggests outward-facing pins at either end of the semicircle, which locate holes in metal fittings on the upper face of the horizon ring. How this accommodated the meridian ring is not explained, but it seems likely that these fittings sat outside the ring, in line with its eastern face. Because of the hour ring, the semicircle would still have to have been removed and reset when crossing the meridian. Such an arrangement would have the disadvantage of taking the semicircle farther from the surface of the globe.

Do any such fittings or "studs" survive? Are there any holes or filled holes around the north and south points of surviving horizon rings? The celestial globe by Mercator in Amsterdam, with the surviving semicircle, has fragments of metal studs at the north and south points of the horizon ring (*Fig. 10.6*).

44 Moxon, op. cit. (n. 17), p. 7.

FIGURE 10.6 Celestial globe, by Gerard Mercator, 1551; detail of the horizon ring with fragments of metal studs for fixing a semicircle of position (Scheepvaartmuseum, Amsterdam, A.1175(01).
PHOTO: SYLVIA SUMIRA.

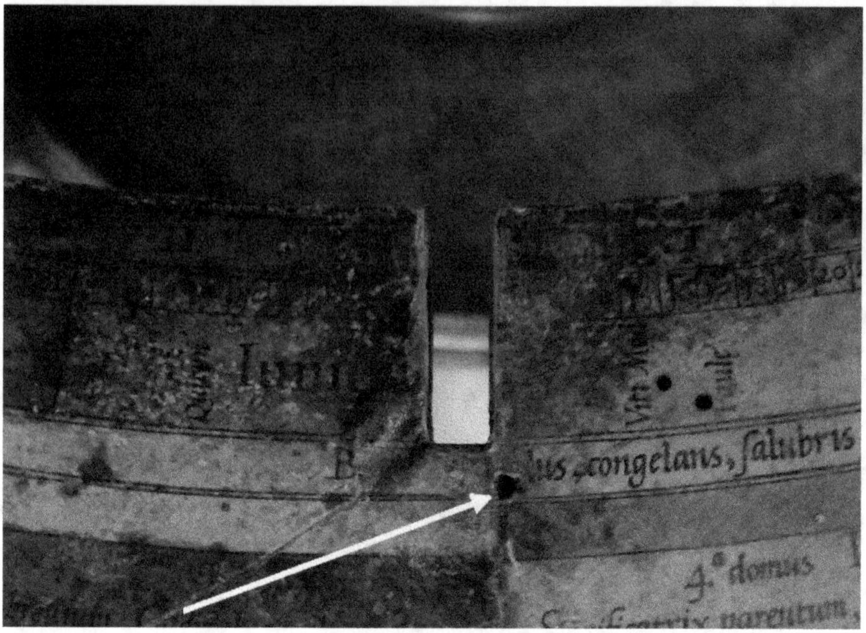

FIGURE 10.7 Celestial globe, by Gerard Mercator, 1551; detail of the horizon ring with holes in a position for studs to fix a semicircle of position (National Maritime Museum, Greenwich, museum no. GLB0097.
PHOTO: SYLVIA SUMIRA.

Because they are broken, they do not show how the semicircle could once have been fitted to them. An example of the same globe (1551, celestial, 42 cm) at the National Maritime Museum, Greenwich, has holes at the appropriate positions on the horizon ring (*Fig. 10.7*).[45]

We have come across no surviving planetary marks added to globes. To learn more about the process and the chances of marks surviving, if the original surface or varnish has survived, we decided to follow Moxon's instructions, using a new globe made from reproductions of gores from *c.* 1600.

Both Blaeu and Moxon write about marking planetary positions on the surface of the globe, but only Moxon gives advice about how the marks should be made and how they can be removed:

45 National Maritime Museum, Greenwich, museum no. GLB0097; Dekker, op. cit. (n. 3), pp. 91–95 and 413–415.

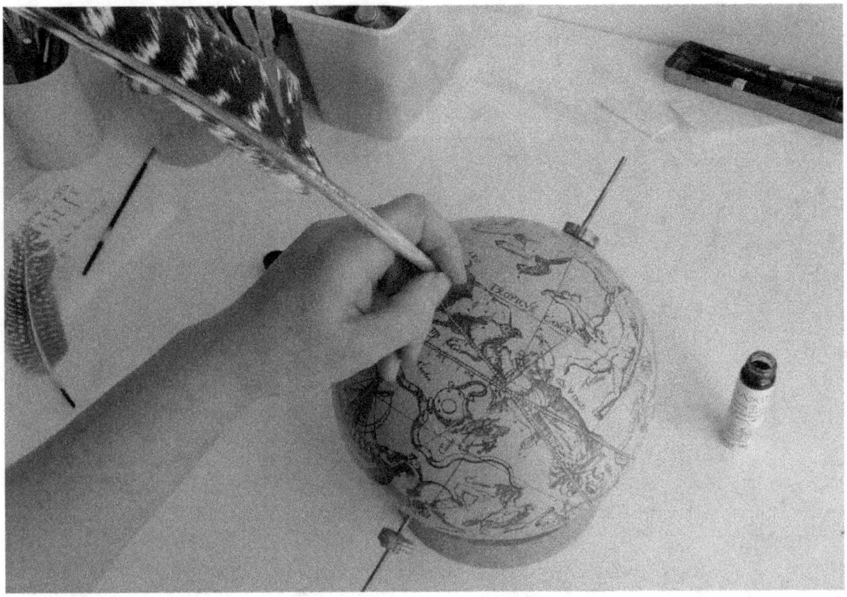

FIGURE 10.8 Applying a planetary mark in ink to the varnished surface of a celestial globe.
PHOTO: JIM BENNETT.

> place a Mark for them on the Globe, either with Ink, if your Globe be Varnisht, for then you may with a wet finger wipe it off again; or with Black-lead, if it be not Varnisht, and then you may rub it out again with a little White Bread [...].[46]

Half of our globe was coated with the natural resin varnish "dammar", and two planetary marks added, representing Venus and Saturn. Iron gall ink, commonly used for writing in the seventeenth century, was applied with a goose quill pen (*Fig. 10.8*). On the unvarnished half, Venus and Saturn were again marked, this time with a soft graphite pencil, graphite being commonly used in the same period, though often referred to as black lead or lead pencil (*Fig. 10.9*).

One day later the marks for Venus, whose position in the sky changes quickly, were removed, with a wet finger and white bread as Moxon advised. Both the ink and the graphite marks came away easily, leaving no discernible trace. The marks for Saturn were left for two weeks, the maximum period suggested by Moxon, after which the graphite mark was fairly easy to remove from the unvarnished surface. The ink mark, however, had become very noticeably

46 Moxon, op. cit. (n. 17), pp. 105–106.

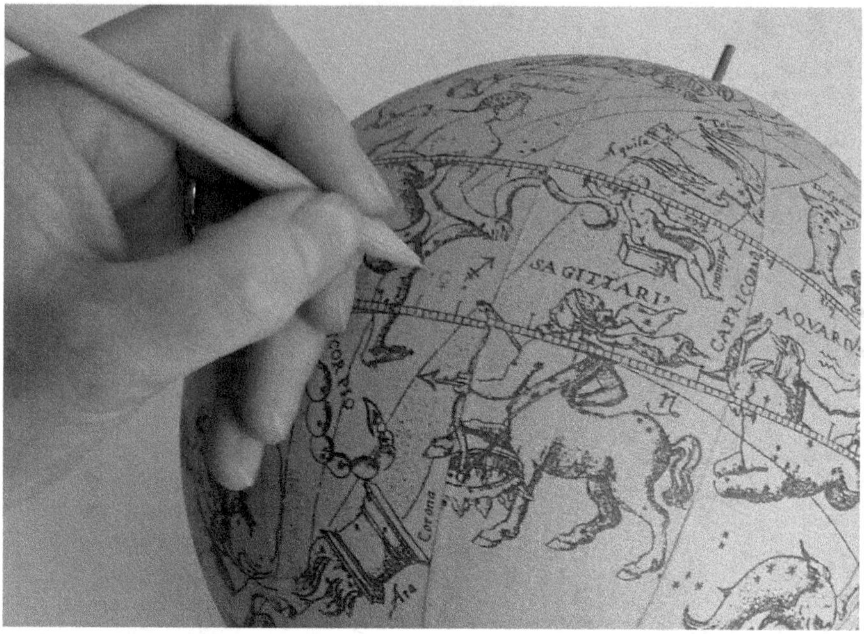

FIGURE 10.9 Applying a planetary mark in pencil to the unvarnished surface of a celestial globe.
PHOTO: JIM BENNETT.

darker, having developed the rich black tone characteristic of iron gall ink. It was all but impossible to remove and persistent attempts rubbed off the varnish, leaving an inky smudge.

If ink marks are found on old varnish in positions that could coincide with planetary co-ordinates, they might be evidence of astrological use. These would fall within the zodiacal band, though Moxon also mentions the possibility of marking the position of a comet. We know of no such marks but have not searched in any systematic way. If such marks are ever found on a varnished surface in the course of conservation, when they are perhaps most likely to be noticed, they would present an interesting ethical dilemma.

Conclusion

Our conclusions are modest. What we have seen, taken together, constitutes a reasonable amount of evidence for the astrological use of the "ordinary" celestial globe. Early treatises commonly illustrate and describe the circle or semicircle of position, and explain its use for finding the boundaries of the houses of heaven, an entirely astrological concept. Some treatises go as far as to explain how the globe might be used to allocate the planets, the lunar nodes and the Part of Fortune to their respective houses. A few globes survive with a semicircle of position; others retain evidence of having had mounting arrangements for such a semicircle. Since we have not conducted a systematic search and since accessories are not always recognised for what they are, further examples of "material" evidence may emerge.

The evidence does not allow us to say that the globe was commonly used in everyday astrological practice. The strongest instance we have at present remains the letter from Lilly to Ashmole: "I set my figure by the Globe". Even so, when Lilly wrote a practical, vernacular handbook, "a most easie Introduction to the whole Art of Astrology", namely his *Christian Astrology* of 1647, he went through the entire process of erecting a figure using tables and without mentioning a globe, which would have been beyond the financial means of most of his readers.[47]

---

47 William Lilly, *Christian Astrology*, London, Thomas Budenell, 1647, pp. 33–42.

CHAPTER 11

Instruments and the Astrologer's Image

Richard Dunn

This chapter is a provisional exploration of visual and literary imagery as evidence of the role of instruments in astrology.* Instruments such as astrolabes and globes can be used for astrological functions, as is discussed elsewhere in this volume. Images of astrologers also often show a range of instrumental and other aids. Yet there remains a question of the extent to which practising astrologers really did make use of scientific instruments in routine practice. The discussion in this chapter focuses on European representations in the period 1600–1750. In particular, it looks at evidence concerning the work of English consultant astrologers including Simon Forman, Richard Napier and William Lilly to think about how imagery might relate to the realities of practice.

A number of issues arise in exploring this subject. The first concerns terminology. Throughout the period under discussion, terms such as astrologer and astronomer could be used in a range of ways, sometimes synonymously, at others for the sake of contrast (e.g. in attacks on astrology). For this chapter, the images chosen are examples that specifically identify the subject as an astrologer or where the context is explicitly astrological. A comparison with representations of astronomers, which often closely resemble those of astrologers, could also be illuminating, but does not lie within the scope of this initial examination.[1]

It is also important to be clear about the different forms of astrological practice. Patrick Curry provides a useful categorization into three broad areas: popular, judicial and high or elite astrology.[2] Between them, these cover a huge range of practices, from those of a rural practitioner drawing on astrological,

* I would particularly like to thank Lauren Kassell and Rob Ralley of the Casebooks Project (www.magicandmedicine.hps.cam.ac.uk), who directed me to many of the specific examples cited.

1 For example, "The Astronomer", by Jusepe de Ribera, oil on canvas, 1638, Worcester Art Gallery, inv. no. 1925.116; "Astronomer by Candlelight", by Gerrit Dou, oil on panel, late 1650s, The J. Paul Getty Museum, Los Angeles, inv. no. 86.PB.732; "The Astronomer" (which has been known as "The Astrologer"), by Johannes Vermeer, oil on canvas, 1668, Musée du Louvre, Paris, inv. no. 1983–28.

2 Patrick Curry, *Prophecy and Power: Astrology in Early Modern England*, Cambridge, Polity Press, 1989, pp. 7–15.

 | DOI:10.1163/9789004381438_013

magical and other beliefs as part of a rich therapeutic mix, through authors of almanacs and astrological consultants such as those discussed below, to those writing more philosophically about the nature and operation of astrological influences. This chapter concentrates on judicial astrology, which itself ranges from making predictions about an individual based on the planets' positions at the moment of birth, to judging the likely outcome of some enterprise (electional astrology), to horary astrology, which sought to answer a question based on the map of the heavens at the time the question was asked.[3] A key tool for judicial astrology was the figure, a chart showing the positions of the planets and other astrological parameters at a specific time. Most of the working examples discussed later in this chapter comprise the casting and interpretation of horary figures.

This broad characterization suggests a number of contexts and ways in which instruments might be used: fundamental observations to produce the data and tables needed to draw up an astrological figure; determining the time of a specific event as the basis of a forecast (e.g. time of birth or at which a question is asked); and marking a moment already determined as propitious for beginning a particular action or endeavour. One can find examples of each in astrological writings. John Dee, for example, refers to the first in the *Monas Hieroglyphica* (1564), when he suggests that the symbolic language he describes will reform observational practice. Will not the "astronomus" (meaning both astrologer and astronomer), he asks,

> be very sorry for the cold he suffered under the open sky, for [all his] vigils and labours, when here, with no discomfort to be suffered from the air, he may most exactly observe with his eyes the orbits of the heavenly bodies under [his own] roof, with windows and doors shut on all sides, at any given time, and without any mechanical instruments made of wood and brass?[4]

Claudius Ptolemy's *Tetrabiblos*, which was published in many editions from the fifteenth century onwards and became the major reference work for astrological practitioners, contains an example of the second usage:

3 *Ibid.*, p. 8; Keith Thomas, *Religion and the Decline of Magic*, London, Weidenfeld and Nicolson, 1971, pp. 284–288.

4 John Dee, *Monas Hieroglyphica*, Antwerp, Willem Silvius, 1564, p. 6r, translation from Conrad H. Josten, "A Translation of John Dee's *Monas Hieroglyphica* (Antwerp, 1564), with an Introduction and Annotations", *Ambix* 12 (1962), pp. 84–221: 131.

> Difficulty often arises with regard to the first and most important fact, that is, the fraction of the hour of the birth; for in general only observation by means of horoscopic astrolabes at the time of birth can for scientific observers give the minute of the hour, while practically all other horoscopic instruments on which the majority of the more careful practitioners rely are frequently capable of error, the solar instruments by the occasional shifting of their positions or of their gnomons, and the water clocks by stoppages and irregularities in the flow of the water from different causes and by mere chance.[5]

Finally, Luca Gaurico provides an example of the use of instruments as part of electional astrology in his *Tractatus astrologicus* (1552), recalling that he,

> computed the hour and erected the celestial figure for the laying of the first stone of the foundation of the building near St Peter's [new Farnese wing of the Vatican, April 1543]. But Vincenzo Campanazzi of Bologna used an astrolabe to determine the suitability of the time, crying out with a loud voice, "Behold, the sixteenth hour from the accustomed setting of the clock is now complete." And at once Ennio of Veroli, cardinal of Albano, having put on a white stole and wearing a cardinal's tiara on his head, fitted into the foundation a great piece of marble, beautifully polished and adorned with the arms of Pope Paul III.[6]

Astrological Iconography

So what of the pictorial representation of astrologers? For the seventeenth century, at least, one place to start is with images in books of trades, such as Jan and Casper Luyken's *Het Menselyk Bedryf* (The Human Trades), printed in several editions from 1694 onwards (*Fig. 11.1*). The Luykens' book followed the tradition of earlier works, such as the *Eygentliche Beschreibung Aller Stände*

5 Claudius Ptolemy, *Tetrabiblos*, ed. and trans. by Frank Egleston Robbins, "Loeb Classical Library 435", Cambridge MA, Harvard University Press, 1940 (reprint: London, William Heinemann Ltd., 1980), pp. 229–231. Jacobus Rueff, *De Conceptu, et Generatione Hominis*, Frankfurt, S. Feyerabend, 1580, book 1, p. 3r, has an illustration in which two astrologers watch the heavens to cast a figure as a woman gives birth. Dividers are the only instruments shown.

6 Luca Gaurico, *Tractatus astrologicus*, Venice, B. Caesanus, 1552, p. 7r; translation in Anthony Grafton, *Cardano's Cosmos: The Worlds and Works of a Renaissance Astrologer*, Cambridge MA, Harvard University Press, 1999, p. 100.

MENSELYK BEDRYF. 177

DE ASTROLOGIST.

Daar 't meeſt aan is gelegen, Staat meeſt te overwegen.

Zo laag in 't ſtof te zyn gezeten,
En 's hoogen hemels loop te meeten,
Schynt veel: maar 't is van veel meer nut,
Den loop des levens naar te ſpeuren,
En wat 'er eind'lyk ſtaat te beuren,
Op dat men 't eeuwig onheil ſchutt'.

M 5 MAT-

FIGURE 11.1 "De Astrologist"; from Jan and Casper Luyken, *Het Menselyk Bedryf*, Amsterdam, Kornelis vander Sys, 1718 (Private collection). PHOTO © NATIONAL MARITIME MUSEUM, GREENWICH.

auff Erden (Exact Description of All Ranks on Earth; Frankfurt, 1568) by Hans Sachs and Jost Amman. These essentially moral texts presented illustrations of different occupations, each accompanied by a short commentary in verse.[7] *Het Menselyk Bedryf* follows this format, with "De Astrologist" showing a solitary figure in a study with a view to the sky. Sitting on the table are a globe and armillary sphere, with a cross-staff lying on the floor and books and papers around the astrologer.

In its depiction of solitary contemplation, the Luykens' astrologer seems to draw on the depiction of St Jerome in his study, which was well established by the sixteenth century as the model for the lone scholar.[8] A contrast is provided in another book of trades, Christoph Weigel's *Centi-folium stultorum in quarto, oder, Hundert ausbündige Narren in Folio* (1709), in that the astrologer appears with clients (*Fig. 11.2*). The querent is shown paying the consultant, who points towards zodiacal and planetary symbols. Behind the astrologer are various tools of the trade: books, a celestial globe, dividers, a telescope, and what appears to be a lens. In both images there are items that commonly appear in representations of astrologers and which symbolize the activities on which their practice might draw: celestial observation (cross-staff and telescope); representation and measurement (globe, armillary sphere, dividers); codification and calculation (books and papers).

The instruments depicted reflect well-established iconographies. Dividers can represent the disciplines of astronomy and geometry, professions such as navigation and architecture, qualities such as justice, temperance and prudence, and acts of creation.[9] Globes can be a symbol of the heavens, but can have more ambivalent meanings. As Catherine Hofmann has written:

> On the one hand, the rotating sphere evokes the fickleness of love and the reverses of Fortune, but also the world doomed to impermanence and death, along with the vanity of all earthly things. On the other, as

7 Alison M. Kettering, "Men at Work in Dutch Art, or Keeping One's Nose to the Grindstone", *The Art Bulletin* 89, 4 (2009), pp. 694–714. Amman's woodcuts include an "Astronomus" (astrologer), shown alone in a study with globes, dividers and sundials.

8 Anna Jasińska, "The Portrait of a Scholar in Painting. An Outline of the History and Development of Iconographic Types", in A. Jasińska, E. Wyka (eds.), *Uczony I Jego Pracownia. The Scholar and his Study*, Kraków, Jagiellonian University Museum, 2005, pp. 9–23.

9 James Hall, *Dictionary of Subjects and Symbols in Art*, revised edition, London, John Murray, 1979, p. 73; Juan Eduardo Clirlot, *A Dictionary of Symbols*, Mineola, NY, Dover Publications, 2002, p. 61; see also John Block Friedman, "The Architect's Compass in Creation Miniatures of the Later Middle Ages", *Traditio* 30 (1974), pp. 419–429.

FIGURE 11.2 "Astrologischer oder Nativität-Narr"; from Christoph Weigel, *Centi-folium stultorum in quarto, oder, Hundert ausbündige Narren in Folio*, Vienna, Johann Carl Megerle and Nuremberg, Johann Christoph Weigel, 1709, image preceding p. 5 (Getty Research Institute).

> the perfect geometric figure, the globe evokes the power of the mighty or the knowledge of the learned, but also the ultimate heavenly refuge, the firmament that encompasses everything and is supposedly perfect and immutable.[10]

In some representations, the symbolism of globes and spheres relates to the question of astrological determinism. In particular, an image showing a sphere in the palm of the hand suggests that free will and wisdom can counteract or overcome the influence of the stars or fate. This is the reading of such an image in Sebastián de Covarrubias Horozco's *Emblemas morales* (1610; *Fig. 11.3*). Its title, 'Fato Prudentia Maior' (prudence is greater than fate), was a common motto in books of emblems and other works.[11]

Strikingly similar to those of the Luykens and Weigel is another image of an astrologer with a client, from Robert Fludd's *Utriusque cosmi maioris scilicet et minoris metaphysica, physica atque technica historia* (1617–1621), a long and copiously illustrated philosophical discussion of the macrocosm and microcosm (*Fig. 11.4*). The celestial globe and dividers are placed prominently between client and astrologer, who is drawing up an astrological figure. A presumably related text and a pair of spectacles are on the table behind the client, with other books on a shelf. Again, this imagined scene draws on established iconographies.[12]

The instruments these and other authors and artists associate with their astrologers match those found in representations of Astrologia or Astronomia as muses or personifications of the liberal arts. An image from such a series by Georg Pencz in 1541, for example, shows Astrologia and a putto with globe, dividers and a quadrant; a similar image published by Hieronymous Cock in 1565 has a globe, dividers and sundials.[13] Well into the eighteenth century, then, images of astrological practitioners continued to draw on the attributes already well established in the sixteenth for the representation of astrology, astronomy and the scholar.

---

10 Catherine Hofmann, "The Globe as Symbol in Emblem Books in the West, Sixteenth and Seventeenth Centuries", *Globe Studies* 49/50 (2002), pp. 81–120: 119–120.

11 *Ibid.*, p. 105.

12 Curry, op. cit. (n. 2), jacket illustration, and Grafton, op. cit. (n. 6), p. 23, use this image to show the astrological practitioner at work, although they are presumably aware of its fictionalized representation.

13 "Astrologia", from "The Liberal Arts", by Georg Pencz, engraving, about 1541, British Museum, no. E,4.280; "Astrologia", from "Seven Liberal Arts", by Cornelis Cort, after Frans Floris, engraving, published by Hieronymous Cock, 1565, British Museum, no. F,1.292.

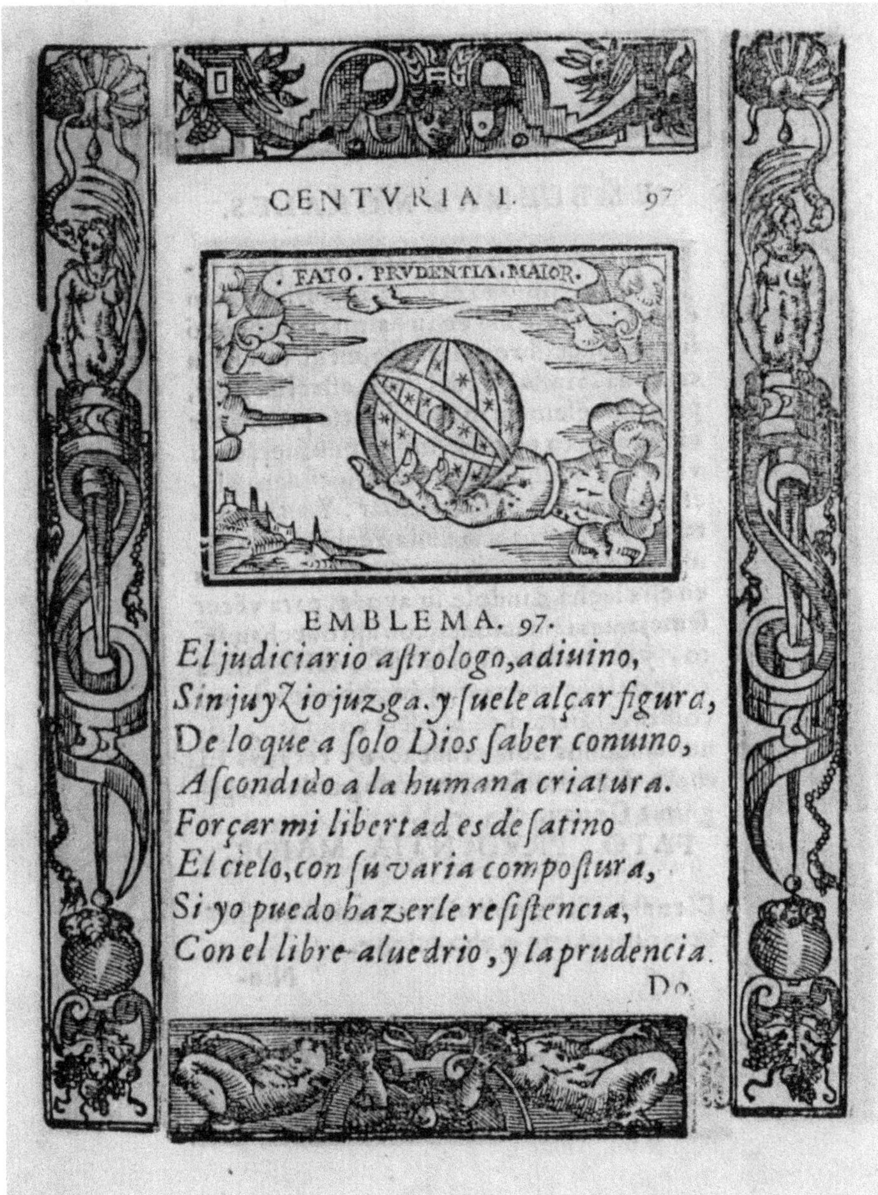
CENTVRIA I. 97

FATO . PRVDENTIA . MAIOR.

EMBLEMA. 97.

El judiciario aſtrologo, adiuino,
Sin juyzio juzga, y ſuele alçar figura,
De lo que a ſolo Dios ſaber conuino,
Aſcondido a la humana criatura.
Forçar mi libertad es deſatino
El cielo, con ſu varia compoſtura,
Si yo puedo hazerle reſiſtencia,
Con el libre aluedrio, y la prudencia.

Do

FIGURE 11.3 Sebastián de Covarrubias Horozco, *Emblemas morales*, Madrid, Luis Sanchez, 1610, p. 97 (University of Illinois Urbana-Champaign).

FIGURE 11.4 Robert Fludd, *Utriusque cosmi maioris scilicet et minoris metaphysica, physica atque technica historia*, Oppenheim, Johann Theodor de Bry, 1617–1621, 2 vols.; v. 2, book I, section II, part IV, p. 71 (Getty Research Institute).

The Consultant's Image

The images in moral and philosophical works were intended as types rather than known individuals. To think more specifically about the instruments of astrology and their representation, one can look to frontispiece images in astrological works, of which a substantial number appeared in England in the seventeenth century.

Inga Elmqvist Söderlund has observed that frontispieces and title page illustrations, which might be a book's only image in this period, are mainly there to indicate what kind of book it is.[14] In 1617, Bernard Bauhuis wrote that, "At the beginning of the book [...] many people would like to see some engraving [...]. It amuses the reader wonderfully, it attracts the buyer, it decorates the book and it does not add much to the price".[15] In addition, a frontispiece portrait often works to establish the author's legitimacy as an authority on the book's subject.

Such a portrait of William Lilly (1602–1681) offers a good example (*Fig. 11.5*). It appeared as the frontispiece to his *Christian Astrology* (1647), the first major textbook of judicial astrology in the English language.[16] Lilly is shown in a gentleman's clothes, accompanied by familiar attributes: his right hand on a celestial globe; an horary figure in his left hand; an open book with the symbols of the signs, planets and astrological aspects, next to a pen and inkstand. An obviously symbolic scene is in the background: Saturn and Jupiter in conjunction near the cusp of Pisces and Aries, above a countryside view with a shepherd and fisherman. This exterior scene draws on Lilly's life. Jupiter and Saturn are in the positions they had when in conjunction in 1643, from which Lilly predicted the downfall of King Charles I. Their positions also reflect the current belief that the birth of Christ had been heralded by a Jupiter-Saturn conjunction on the Aries-Pisces cusp. The pastoral image recalls Lilly's childhood as a farmer's son and suggests a Christian reading of the figures as "Fisher of Souls" and "Shepherd of Men".[17]

The figure in Lilly's hand is significant too. It shows the twelve astrological houses, drawn as triangles within a square, the conventional graphic form for an horary figure at this time, although it lacks the specific positions of the planets and house divisions. Lilly's age is at the top ("Aetatis 45"), while the words "Non cogunt" (they do not compel) are in the middle. This phrase appears in every portrait of Lilly as a statement of his position on astrological determinism: that the stars incline but do not compel. This was the stance adopted by

14 Inga Elmqvist Söderlund, *Taking Possession of Astronomy: Frontispieces and Illustrated Title Pages in 17th-Century Books on Astronomy*, Stockholm, Center for History of Science, Royal Swedish Academy of Sciences, 2010, pp. 88 and 313.

15 Bernard Bauhuis to B. Moretus, 1 August 1617, quoted in *ibid.*, p. 89.

16 Patrick Curry, "Lilly, William (1602–1681)", *Oxford Dictionary of National Biography*, Oxford University Press, 2004, online edition (hereafter *ODNB*), www.oxforddnb.com/view/article/16661 (accessed 28 October 2015).

17 Catherine Blackledge, *The Man Who Saw the Future: A Biography of William Lilly*, London, Watkins, 2015, pp. 47–50.

FIGURE 11.5 William Lilly, engraving by William Marshall; from *Christian Astrology*, London, Thomas Budenell, 1647, frontispiece (Private collection). PHOTO © NATIONAL MARITIME MUSEUM, GREENWICH.

most practising astrologers and was an important statement in the battle to demonstrate that their art did not contradict Christian doctrine, notably the principle of human free will.[18] Indeed, the iconography of Lilly's portrait supports this reading in other ways. Above all, the placing of his right hand on the celestial globe seems to draw on representations of a sphere held in the hand and the notion that the wise man can overrule the stars.

Similar portraits appeared in other English astrological works. Frances Wilmoth, for instance, has noted the close resemblance of different portraits of astrologers executed by Thomas Cross.[19] One of these is of Lilly's amanuensis, Henry Coley (1633–1707). He is shown aged 35 in the frontispiece to his *Clavis Astrologiae; or, a Key to the whole Art of Astrologie* (1669), with a celestial globe and dividers (*Fig. 11.6*).[20] Here, however, the dividers are in Coley's hand and he uses them to measure or draw an astrological diagram.

Given that these images have a specific role with respect to the books in which they appear, the question is whether they reveal anything about day-to-day practice. Frontispiece portraits were not, as these examples show, documentary-style representations of the astrologer in his working environment. It is possible, however, to draw on other sources which help reveal further details of the consultation.

One body of evidence comes from the surviving casebooks of English consultants including Lilly, Simon Forman (1552–1611), Richard Napier (1559–1634) and John Booker (1602–1667). These indicate that clients typically consulted the astrologer in person or perhaps posed questions by letter or through some intermediary. The consultation took place in a study or similar room at the astrologer's own house. William Lilly had a "Corner House on the Strand" in London, while Henry Coley's practice was at his home in Baldwin's Gardens, off Gray's Inn Road, also in London. Forman occupied a number of residences in or near the city during his life, using a study for his consultations and other work, although he did go to patients' houses on some occasions. Like Forman,

18 Ann Geneva, *Astrology and the Seventeenth Century Mind: William Lilly and the Language of the Stars*, Manchester, Manchester University Press, 1995, pp. 9–10; Thomas, op. cit. (n. 3), pp. 361–364.

19 Frances Wilmoth, *Sir Jonas Moore: Practical Mathematics and Restoration Science*, Woodbridge, Boydell Press, 1993, pp. 66–68.

20 This example is unsigned, but another in the Victoria and Albert Museum, London (museum no. E.275–1960) has the inscription "Cross -sc". For comparison, see the frontispiece by Cross for John Gadbury, *Natura Prodigiorum*, London, J.C. for F. Cossinet and T. Basset, 1660. For Coley, see Bernard Capp, "Coley, Henry (1633–1704)", *ODNB*, www.oxforddnb.com/view/article/5899 (accessed 8 November 2015); Curry, op. cit. (n. 2), pp. 88–89.

FIGURE 11.6 Henry Coley, engraving by Thomas Cross; from *Clavis Astrologiae; or, a Key to the whole Art of Astrologie*, London, Printed for Joshua Coniers, 1669, frontispiece (Private collection).
PHOTO © NATIONAL MARITIME MUSEUM, GREENWICH.

John Booker operated from addresses in London, while Richard Napier worked from his rectory in Great Linford in Buckinghamshire.[21]

On receiving the question, the astrologer recorded the time at which it was asked, drew up a figure for that time and used it to frame a judgement or forecast. Experienced practitioners might be able to carry out a full consultation in quarter of an hour or less. The numbers seen were impressive, therefore. Forman recorded over 10,000 consultations between 1596 and 1603, Napier 70,000 between 1597 and 1634, and Booker more than 16,500 between 1648 and 1665.[22]

Establishing specific details of the role of instruments and other material aids is more difficult, however, since casebook entries usually show at most the question, figure and judgement.[23] Based on his reading of the casebooks, however, Michael MacDonald has reconstructed Napier's routine:

> When the precise coordinates of the patient's identity had been recorded, the astrologer noted the time and date and drew a cross-hatched box [i.e. the horary figure] on the page. Deciphering the ephemeris at his elbow, he mapped the heavens at that moment, positioning the symbols of the planets and the signs of the zodiac along the celestial frame. The purpose of this astral cartography was to situate the patient in the cosmos, placing him at the vortex of the natural forces that impelled the universe, discovering the correspondences that linked microcosm and macrocosm.[24]

The main material element MacDonald identifies is the casebook containing the horary figures and notes. Kassell agrees in her discussion of Forman's practice, noting that the casebook was the main thing present at consultations.

21 Jonathan Andrews, "Napier, Richard (1559–1634)", *ODNB*, www.oxforddnb.com/view/article/19763 (accessed 28 October 2015); Bernard Capp, "Booker, John (1602–1667)", *ODNB*, www.oxforddnb.com/view/article/2865 (accessed 28 October 2015); Capp, op. cit. (n. 20); Curry, op. cit. (n. 2), pp. 29 and 88–89; Curry, op. cit. (n. 16); Lauren Kassell, "Forman, Simon (1552–1611)", *ODNB*, www.oxforddnb.com/view/article/9884 (accessed 28 October 2015); Lauren Kassell, *Medicine and Magic in Elizabethan London: Simon Forman: Astrologer, Alchemist and Physician*, Oxford, Clarendon Press, 2005, p. 137.

22 Casebooks Project, "Introduction to the casebooks" www.magicandmedicine.hps.cam.ac.uk/the-manuscripts/introduction-to-the-casebooks (accessed 28 October 2015); Curry, op. cit. (n. 2), pp. 8–9; Kassell, "Forman", cit. (n. 21); Thomas, op. cit. (n. 3), pp. 305–307.

23 Casebooks Project, "How to read the casebooks", www.magicandmedicine.hps.cam.ac.uk/the-manuscripts/how-to-read-the-casebooks (accessed 28 October 2015).

24 Michael MacDonald, *Mystical Bedlam: Madness, Anxiety, and Healing in Seventeenth-Century England*, Cambridge, Cambridge University Press, 1981, p. 26.

"Astrology", she emphasizes, "was by definition a written art".[25] MacDonald, indeed, suggests that it was the horary figure above all that was the fundamental difference between astrological and other medical practitioners.[26]

The other aid MacDonald identifies is the ephemeris, the astronomical tables of data needed to construct the horary figure. As Lilly noted in his guide to judicial astrology, "without a present *Ephemeris* and *Table of Houses* it's impossible to instruct you to set a figure, without which we can give no judgement, or perform anything in this art".[27] There is evidence of the ephemerides Napier and Forman used. Napier seems to have used more than one and sometimes compared data from different ephemerides. Casting a figure on 15 May 1599, for example, he includes data from "Everart" and "Stadius". Likewise, "Orig[anus]" and "Everart" are identified as sources in January 1602.[28] In another example, his judgment seems to hinge on which he chooses to trust in diagnosing a possible pregnancy: one by Everart, the other not identified.[29] Forman also refers to Everart in several cases, and claimed during examination by the College of Physicians in 1593 that he practised his art "using no help but the Ephemerides".[30]

There is some evidence of paper-based material aids, therefore, but what of instruments? Here the evidence becomes frustratingly scarce, with Forman's statement to the College of Physicians suggesting that instruments were of only marginal use to him. One exception occurs in Richard Napier's casebooks. In July 1598, he cast a figure on behalf of Harry Mun, who had sent a urine sample. Napier noted that, "Mun sent his water to me about [...] 6 [...] but we

25 Kassell, *Medicine and Magic*, cit. (n. 21), pp. 129 and 132.

26 MacDonald, op. cit. (n. 24), p. 30.

27 William Lilly, *Christian Astrology*, London, Thomas Budenell, 1647, p. 32.

28 Richard Napier, "Figure for William Colles", 15 May 1599: Bodleian Library, Oxford, MS Ashmole 228, fol. 164r, www.magicandmedicine.hps.cam.ac.uk/view/case/normalised/CASE12464 (accessed 28 October 2015); Napier, "Figure for Mrs Garnar", 2 January 1602: Bodleian Library, Oxford, MS Ashmole 404, fol. 256v, www.magicandmedicine.hps.cam.ac.uk/view/case/normalised/CASE20187 (accessed 28 October 2015). These presumably refer to Martin Everart, *Ephemerides novae et exactae*, Leiden, 1597 and later editions; Johannes Stadius, *Ephemerides*, Lyon, 1585 and other editions; David Origanus, *Ephemerides novae*, Frankfurt an der Oder, A. Eichorn, 1599.

29 Napier, "Figure for Mrs Troughton's servant", 2 January 1602: Bodleian Library, Oxford, MS Ashmole 404, fols. 256v–257r, www.magicandmedicine.hps.cam.ac.uk/view/case/normalised/CASE20189 (accessed 28 October 2015).

30 For example, Simon Forman, "Figure for himself", 1 January 1599: Bodleian Library, Oxford, MS Ashmole 219, fols. 135r–135v, www.magicandmedicine.hps.cam.ac.uk/view/case/normalised/CASE5888 (accessed 28 October 2015). Forman's evidence to the College of Physicians is quoted in S.P. Cerasano, "Philip Henslowe, Simon Forman, and the Theatrical Community of the 1590s", *Shakespeare Quarterly* 44, 2 (1993), pp. 145–158: 146.

aymed the houre because it was rayny".[31] This suggests that Napier was using a sundial or similar instrument for determining the time, but was unable to do so due to the inclement English weather and had to estimate (aim) instead. It does seem likely, however, that most consultants would have used sundials for timekeeping: they were cheaper and, weather permitting, more reliable than clocks and watches. As Sylvia Sumira and Jim Bennett note in their chapter in this volume, William Lilly's correspondence with Elias Ashmole contains another interesting reference. Writing in December 1664 about a comet, the position of which he was keen to pinpoint, Lilly notes that, "I set my figure by the Globe", but was unable to fix its position precisely, "having no Compass".[32] This is not part of a consultation, but points to an additional use of instruments in an astrological context.

The primary evidence of the work of English consultant astrologers tentatively suggests, therefore, that the casebook containing the astrological figure and consultation notes was the key material aid, while an ephemeris was a secondary aid, used in drawing up the figure. Instruments for time-telling were used, but there is little evidence of a regular operational role for globes, dividers or astrolabes.[33] Although further examination of surviving casebooks may shed more light on this matter, the question remains as to whether instruments were routinely present and, if so, what role they had.

The Eye of the Beholder

One can look to other sources to tackle this question of what other aids an astrological consultant might have had in the room. For seventeenth-century London, the most striking description is in John Melton's *Astrologaster, Or, The Figure-Caster* (1620). This must be treated with caution, however, since the work is an attack on false practitioners, typified in the subtitle as "Artlesse Astrologers, and Fortune-tellers, that cheat many ignorant people under the pretence of foretelling things to come". Nevertheless, Melton's description of an astrologer's study, which is depicted on the title page, is evocative. Visiting

31 Napier, "Figure for Harry [Henry] Mun", 24 July 1598: Bodleian Library, Oxford, MS Ashmole 228, fol. 4v, www.magicandmedicine.hps.cam.ac.uk/view/case/normalised/CASE11572 (accessed 28 October 2015).

32 William Lilly to Elias Ashmole, 29 December 1664: Bodleian Library, Oxford, MS Ashmole 423, fols. 248r–249v, in C.H. Josten (ed.), *Elias Ashmole (1617–1692): His Autobiographical and Historical Notes, his Correspondence and Other Contemporary Sources Relating to his Life and Work*, Oxford, Clarendon Press, 1966, 5 vols.; v. 3, pp. 994–995: 995.

33 Astrolabes, of course, were less commonly used in England by this time.

a "Doctor P.C.", whom he already considers a trickster, Melton describes the study in which the "Wise-man" receives him:

> Before a Square Table, covered with a greene Carpet, on which lay a huge Booke in *Folio*, wide open, full of strange Characters, such as the *Aegyptians* and *Chaldeans* were never guiltie of; not farre from that, a silver Wand, a Surplus, a Watering Pot, with all the superstitious or rather fayned Instruments of his cousening Art. And to put a fairer colour on his black and foule Science, on his head hee had a foure-cornered Cap, on his back a faire Gowne (but made of a strange fashion) in his right hand he held an Astrolabe, in his left a Mathematicall Glasse.[34]

Melton identifies a variety of material and instrumental aids. Books have a prominent place, the astrologer declaring that, "this Booke (meaning his *Ephemerides*) with my Art and Industrie, shall be the Instrumentall Causes to make you happie in the recoverie of that which is worthie both of my Care and your Cost"—Melton has come under the pretence that he has lost a gold chain.[35] Other aids, such as the silver wand and watering pot, have a distinctly magical aura. The "Mathematicall Glasse" is clearly inspired by the recently invented telescope, but is conspicuously more powerful and seemingly mystical in its operation: "with this Instrument", the astrologer claims, "first devised by that learned Man in our Art, *Hermes Trismegistus*, otherwise known as *Mercurie*, I can see all things done in Christendome".[36] Other instruments are more obviously for celestial observation or calculation, notably the astrolabe. Melton, however, damns the astrologer's instrumental aids as "meere superstitious Ornaments", likening them to the trappings of papists (*Astrologaster* is vehemently anti-Catholic).

Thomas Tomkis's play *Albumazar* (1615) also attacked astrology and its false practitioners. In a tale of love and marriage, the astrologer Albumazar, who claims to practise alchemy and necromancy as well, is a fraud who bamboozles his intended victims with verbose gibberish, including the wonderfully overblown compound "Necro-puro-geo-hydro-cheiro-coscinomancy". Albumazar's aids include an ephemeris and an increasingly fantastic set of instruments. These range from the "perspicill" (telescope), to an "Otacousticon"

34 John Melton, *Astrologaster, Or, The Figure-Caster*, London, Barnard Alsop, 1620, p. 8.

35 *Ibid.*, p. 9.

36 *Ibid.*, p. 10.

for improved hearing, which turns out to be an artificial pair of ass's ears, and a "wind instrument" that allows one to speak in ten languages.[37]

In Melton's and Tomkis's descriptions, the role of instruments such as astrolabes and telescopes, alongside other objects, places them as artefacts that help give the (deceiving) astrologer's work plausibility as well as mystery. Their presence assures the client that the practitioner has the right tools for the job—in this case instruments for measuring and studying the heavens—and knows how to use them. Thinking about images of known practitioners such as Lilly and Coley, one can imagine a globe taking on this role alongside a didactic one; for instance, in helping to explain the principles or specifics of a case. There is clearly theatricality in using instruments and other aids in these ways. Indeed, this theatricality is played up and made suspicious by Melton and Tomkis, who suggest that props and the long explanations that accompany them are deliberately used to befuddle the naive client. Nevertheless, one can plausibly imagine a consultant like Lilly deploying instruments in a broadly similar, but not deceitful, way (although perhaps without the exaggerated cap and silver wand).

The use of instruments in attacks on astrologers shows, however, that this may be a double-edged sword, since perception is everything. A suspicious observer might interpret such items as the trappings of a dishonest practitioner or of a magical (therefore suspect) one. This is the line taken by Melton and Tomkis. In other words, the tools an astrologer uses might raise suspicions for those already worried that something is amiss, in particular given the often close and/or generally assumed links between astrology, magic and other arts in the seventeenth century.[38] Some associations could be dangerous.

With this in mind, it is no surprise that images of astrologers in satirical contexts resemble those meant as serious. Effective satire, after all, trades on plausibility. In the second part of *Hudibras* (1664–1678), a mock-epic poem offering a satirical commentary on the English Civil War, Samuel Butler introduces an astrologer, Sidrophel, who parodies William Lilly.[39] Sixty years later the poem was still popular and was illustrated in a series of prints by William Hogarth that appeared in many subsequent editions (*Fig. 11.7*).[40]

37 Thomas Tomkis, *Albumazar. A Comedy* [*1615*], ed. by Hugh G. Dick, Berkeley and Los Angeles, University of California Press, 1944, pp. 80–82, 85 and 96.

38 Thomas, op. cit. (n. 3), pp. 631–635.

39 Curry, op. cit. (n. 16).

40 Mark Hallett, Christine Riding, *Hogarth*, London, Tate Publishing, 2006, p. 64.

FIGURE 11.7 Hudibras and his servant Rolpho with Sidrophel and his assistant Wachum, engraving by J. Mynde, after William Hogarth; from Samuel Butler, *Hudibras*, London, J. Bettenham, 1744, 2 vols.; vol. 2, plate XI (Private collection). PHOTO © NATIONAL MARITIME MUSEUM, GREENWICH.

Butler's Sidrophel is both dishonest and a fool: he uses instruments poorly and knows little of astronomy or astrology. Outside his house, a "*Mansion* prudently contriv'd", stands an obelisk,

> On which was written, not in words,
> But *Hieroglyphick* Mute of *Birds*,
> Many rare pithy Saws concerning
> The Worth of *Astrologick* Learning:
> From Top of this there hung a *Rope*,
> To which he fasten'd *Telescope*;
> The *Spectacles* with which the *Stars*
> He reads in smallest *Characters*.[41]

Sidrophel's instrumental incompetence is revealed as he mistakes a light hung from a kite for Saturn, which he interprets as an omen of the end of the world. Like the targets of other works by Butler, Sidrophel is a "virtuoso", a dabbler in scientific ideas, full of speculation yet oblivious to the real world and easily fooled. In the era that witnessed the foundation of the Royal Society and reports of the extraordinary things they saw fit to investigate and discuss, virtuosi became stock figures of fun in satirical plays and poems.[42]

Later on, Sidrophel and Hudibras debate the validity of astrology, during which the poem's hero makes fun of the astrologer, asking a nonsense question that includes one of the instrumental attributes (the cross-staff) with which astrologers might be depicted:

> Tell me but what's the nat'ral Cause,
> Why on a *Sign* no *Painter* Draws,
> The *Full-Moon* ever, but the *Half*;
> Resolve that with your *Jacob's Staff* [...][43]

---

41 Samuel Butler, *Hudibras*, London, J. Bettenham, 1744, 2 vols; v. 2, p. 41.

42 Richard Dunn, *The Telescope: A Short History*, London, National Maritime Museum, 2009, pp. 52–54; Roslynn D. Haynes, *From Faust to Strangelove. Representations of the Scientist in Western Literature*, Baltimore and London, Johns Hopkins University Press, 1994, pp. 35–49.

43 Butler, op. cit. (n. 41), p. 73.

The debate becomes a fight, in which Hudibras overpowers Sidrophel and empties the astrologer's pockets,

> And found a *Watch*, with *Rings* and *Lockets*,
> Which had been left with him t'erect
> A *Figure* for, and so detect;
> A *Copper-Plate*, with *Almanacks*
> Engrav'd upon't, with other Knacks,
> Of *Booker's*, *Lilly's*, *Sarah Jimmers*,
> And *Blank-Schemes*, to discover *Nimmers*;
> A *Moon Dial*, with *Napier*'s Bones,
> And several *Constellation* Stones,
> Engrav'd in *Planetary Hours*,
> That over *Mortals* had strange Powers,
> To make 'em thrive in *Law* or *Trade*,
> And Stab or Poison to evade;
> In *Wit* or *Wisdom* to improve,
> And be victorious in *Love*.[44]

The description identifies a range of items a real astrologer might use or have around him: the figure he has been paid to erect; the plate with almanacs on it (the equivalent of an ephemeris); calculation aids (Napier's bones); and a time-telling device (the Moon dial). But Butler also deliberately mixes astrological and magical imagery. Sidrophel's pockets include not just items for making predictions, but also for effecting changes by means of their "strange Powers". Hogarth's illustration (*Fig. 11.7*) draws on familiar imagery too: the astrologer erecting a figure; books on the shelf above; instruments including a globe, telescope and cross-staff on the floor. At the same time, the hanging crocodile and chemical/alchemical apparatus on a shelf add to the deliberately mixed messages.

Another virtuoso/astrologer based on Lilly appears in William Congreve's *Love for Love* (1695). In the play, Sir Sampson Legend believes he is to be married to Angelica, the niece of the astrologer, Foresight, for whom the ephemeris is an important prop. Foresight is also shown with instrumental aids. When Angelica asks her uncle to give her away at the supposed marriage, Sir Sampson interjects, "That he shall, or I'll burn his Globes".[45] It is a rare identification of one of the instruments so often associated with astrological practice in the

44 *Ibid.*, pp. 96–97.

45 William Congreve, *Love for Love. A Comedy*, London, Printed for Jacob Tonson, 1695, p. 60.

visual imagery of consultant and other astrologers. The intent is satirical, but it draws on what must have been plausible elements.

Conclusion: The Rhetoric of Representation

This brief survey has suggested that the visual and literary iconography of the astrological practitioner had a number of stable elements between the sixteenth and eighteenth centuries. These offered a recognizable suite of instruments that helped construct the astrologer's image, but which were susceptible to satirical and polemical treatment. What is less clear is how these images related to the daily practices of astrological consultants like William Lilly.

One suggestion is that some of the instruments found in the consulting room took on a symbolic and didactic rather than operational role. This symbolism mirrored that of visual and literary representations of the astrologer and made similar claims. To take a final example, one can look at an obviously imaginary picture, Nicolas de Larmessin's "Habit d'Astrologue" (*Fig. 11.8*), another from a series of images of trades. The astrologer is adorned with the twelve zodiacal signs, placed on the parts of the body they were thought to govern (thus mimicking the image of "Zodiac Man"). The other attributes make a rhetorical claim about astrology and its foundations. Astrology's legitimacy, they assert, comes from astronomical observations made with instruments: the Observatoire de Paris (only about 30 years old at this time) with a long telescope on the roof is in the background, while the works of Copernicus and Tycho Brahe adorn the astrologer's arms. In addition, he holds a pair of dividers in his right hand, an armillary sphere in his left. The wise and prudent man, this emphasizes, draws on instrumental observations, directly or through their codification in published tables, as a basis for forecasts that allow him and his client to be masters of their own fates. Through the use of instruments and codified knowledge, the astrologer brings a chaotic universe into order.

This, I would suggest, makes some sense of what might have gone on in the seventeenth-century consulting room. The astrologer's key tool (the figure) was written and his key aid (the ephemeris) was printed. Instruments such as sundials had a time-telling function, which was essential for casting the figure on which the judgement depended. Other instruments had a more symbolic and didactic role in the daily routine. So when John Dee suggested that his new system would do away with cold nights on a hillside observing with instruments, he may have been overstating his case. The astrologers, it seems, were already indoors.

FIGURE 11.8 "Habit d'Astrologue", engraving by Nicolas de Larmessin, about 1695 (Private collection).
PHOTO © NATIONAL MARITIME MUSEUM, GREENWICH.

CHAPTER 12

Defining Scientific Instruments in Astrological Practice: A Response

Marvin Bolt

Astrology of some kind appears in virtually every known culture and is highly adaptable to diverse situations, times and places. European astrology is inextricably linked with Ptolemy's *Tetrabiblos*, which became available in Latin through Plato of Tivoli's translation in 1138. By then, numerous commentaries in Arabic, astrological texts and instruments with astronomical and astrological markings had already extended Ptolemaic astrological concepts, informed by new, dynamic religious and intellectual contexts. By the fourteenth century, the literature on astrology had grown considerably, as had controversies about its compatibility with Aristotelian philosophy and Islamic, Jewish and Christian doctrines.

In fact, as many scholars have noted, there were several varieties of astrologies and ways to classify them.[1] One might distinguish between mathematical/astronomical, natural/physical and spiritual astrologies in one scheme, or between popular, judicial and elite astrologies in another. Moreover, these astrologies appeared in diverse contexts, including theology, medicine, politics, civics, meteorology and more ordinary human affairs. All of which is to say that, like love, European astrology in the medieval, Renaissance and early modern periods was a many-splendored thing. Indeed, it was no one thing at all, and in order to understand astrology in any detail, we must direct our attention to local contexts.

The papers in this volume explore a variety of astrological episodes in which scientific instruments appear to have played some role. These papers thereby also provide an opportunity to reflect on the nature and even category of scientific instruments prior to, and during, the emergence of modern science.

1 See, for example, Nicholas Campion, *Introduction to the History of Astrology*, London, Institute for the Study of Cycles in World Affairs, 1982; Roy Willis and Patrick Curry, *Astrology, Science, and Culture*, London, Berg Publishers, 2004; Patrick Curry, "The Historiography of Astrology: A Diagnosis and a Prescription", in Günther Oestmann, H. Darrel Rutkin, Kocku von Stuckrad (eds.), *Horoscopes and History*, Berlin, Walter de Gruyter, 2005, pp. 261–274; Patrick Curry, Angela Voss (eds.), *Seeing with Different Eyes: Essays in Astrology and Divination*, Newcastle, Cambridge Scholars Publishing, 2007.

 | DOI:10.1163/9789004381438_014

Paraphrasing Peter Dear's evocative question posed a decade ago to historians of science, we might ask, what is the history of scientific instruments the history *of*? Or, what is scientific about these and other astrological instruments? Dear suggests that we ought to consider "science" as consisting of two changing, interrelated enterprises, a natural philosophic "discourse of contemplative knowledge" and an instrumental "discourse of practical or useful knowledge".[2] Guided by this distinction, I encourage readers of this volume to consider a series of related issues illustrated to varying extents in these papers:

1. Is the instrument under discussion a standard, canonical form, or more in the nature of a one-off?
2. Is its instrumentality a matter more of computational problem-solving or of rhetorical demonstration?
3. In what way is its instrumentality private or public? If public, what kind of public?
4. What is the role of the local context in the production and use of the instrument? How generalizable is the claim being made? What is the extent, and what are the limits, of what this instrument can tell us?

In the Introduction, the volume editors remind us that instruments do not speak for themselves and that astrological traditions, interpretations and practices change from place to place and from time to time. For example, even something as fundamental as the number of planets can vary, from five to seven or even eight (considering the lunar nodes, a point noted by Günther Oestmann as well). The complexity and variability of astrological doctrines and markings require us to look at specific examples and local contexts, and to examine the particularities of a given text or instrument. In turn, Oestmann directs attention to a specific late-seventeenth-century device he calls a tabular instrument, which provides information and tools for conducting horoscopic work. These data include astrological houses, planetary properties and positions, zodiacal properties, dignities and more, and serve as successors or parallels to the more commonly known *zijes*.

Petra Schmidl takes on the challenge by examining in detail four of the earliest historic texts on astrolabes and three of the earliest known astrolabes. Comparing texts by Ioannes Philoponus and Severus Sebokht, and then later manuscripts by al-Khwārizmī and ʿAlī b. ʿIsā, Schmidl compares and contrasts their features in order to establish their potential roles for astronomical and astrological purposes, making a preliminary distinction between astrological, non-astrological and ambiguous markings. The latter consist of features on

2 Peter Dear, "What is the History of Science the History *of*? Early Modern Roots of the Ideology of Modern Science", *Isis* 96, 3 (2005), pp. 390–406: 390.

astrolabes that could be used for astrological and/or astronomical purposes, but whose purpose remains ambiguous in the absence of further evidence. Explicitly acknowledging such ambiguity is an important aspect of studying these texts and instruments and will be a useful distinction in future discussions. Schmidl notes that in contrast to the earlier texts, the Islamic treatises show a "conspicuous development" of material addressing astrological problems. She alludes to the challenges of incorporating Greek knowledge into a foreign culture and notes the simultaneity of the appearance of astrological content, in ambiguous forms, in the early Abbasid period with the appearance of astronomical content related to daily ritual prayers and obligations. In light of the incompatibility of the Islamic notion of free will with the deterministic aspects of astrology, ambiguity can serve as a useful, deliberate and theologically safe strategy for supporting both astrological and astronomical content.

Unfortunately, the earliest known astrolabes, at least one of which has, sadly, disappeared in recent years, do not have astrological markings, but one example, made by al-Khujandī and dated 374 *hijra* (984–985), includes one unusual plate that is deemed to be solely for astrological purposes. Schimdl also notes varying references on instruments and in texts to planetary names by using the final letter of their name or by using their planetary symbol, and infers that those practices are suggestive of astrological uses. Schmidl's detailed study concludes that astrolabes played a role in timekeeping in late antiquity and gained an astrological purpose when astrolabes made their way east. Nonetheless, Schmidl is rightly concerned about synthesizing evidence from diverse places and times, noting that the astrolabe is highly adaptable, acquiring different purposes as religious and aesthetic sensibilities evolve and as mathematical and technical skills develop.

Giorgio Strano raises similar concerns, noting that they apply even in the case of a single instrument maker, in this case, Antonio Santucci. Looking at one text, a treatise on an armillary sphere, and two physical armillaries (one in Madrid, the other in Florence) based on that work, all three of which were produced between 1582 and 1593, Strano has carried out his trademark archival sleuthing to show significant differences between these manifestations of the same underlying device. After providing a helpful and welcome biographical background of Santucci, Strano compares and contrasts these three objects, emphasizing their contributions to observation, timekeeping, astronomical computations, education and astrology. In addition to clear descriptions of all three objects, what emerge from this careful study are the considerable variations between the text and the two instruments based on it: these manifest different ideas, different styles, differences in funding, different patrons, different knowledge embedded in each one, different principles of dividing celestial houses and a marked increase in technical skills. Those differences, evident in one artisan's variations of the same fundamental device over the course of just a decade, provide compelling reasons to exercise considerable caution about making generalized inferences about scientific instruments in astrological contexts. Moreover, each paper in this volume leads to this conclusion, to which I shall return below.

Strano's paper also lays out the modern principles of conservation that were applied to the Florence armillary, showing how recent and prior conservation efforts have both revealed and concealed information about Santucci's original instruments. This contrasts with Richard Kremer's investigation of a fresco portraying two rather different sundials in Görlitz, a city in southeastern Germany. Here, over a dozen restorations have literally and figuratively blurred its original lines, although with Kremer's keen analysis it is possible to recover at least some of the instrument's original functions, some of which are astrological. Kremer makes a useful distinction between two different kinds of planetary or unequal hours, those that divide the Sun's diurnal path into twelve equal portions, which vary over the course of the year, and those that vary over the course of a single day as well as over the course of a year. This distinction is helpful for Kremer's overall project, an introduction to the appearance of astrological time-keeping in European public spaces from roughly 1350 to 1600. He suggests that whereas contemporary astrological practices were carried out in private spaces, the need for an astrologer to determine time required a departure from that private space and into the outdoor, public arena. For this time period, Kremer lists over a dozen representative public astronomical clocks that also featured unequal hours, concluding that the unequal hours on these likely served a representational function rather than as practical information for time-telling, as it would have been quite difficult for the viewer in the street to read this information. Although found much more commonly on public buildings than on astronomical clocks, sundials rarely featured planetary hours; Kremer notes two exceptions, one of them the dial in Görlitz. He concludes that whereas astronomical time was readily available, and readily accessible, to audiences in public spaces by 1600, astrological time was a rare, yet not unknown, resource. How those time indicators were actually employed remains to be investigated, as does the identification of who required public spaces for astrological time determination.

Marisa Addomine's study of Italian astronomical clocks takes up both questions. She has expanded our awareness of public astronomical clocks in Italy, having located forty of them, in particular one in Mantua, for which we now also have written descriptions of its usage. Her extensive investigation, both in breadth and in depth, provides a useful complement to Strano's work, and yields similar conclusions. Addomine's examination of these clocks reveals a distinctive difference between German and Italian public astronomical clocks: German versions typically include an astrolabic dial, whereas Italian devices feature rotating disks that provide solar and lunar positions, lunar phases, aspects and more. By itself, this result shows evidence of national styles and deserves further investigation.

Her paper pays particular attention to lunar phases and aspects, leaving us looking forward to her future discussion of the distinct ways in which these components appear on various clocks and of additional details about how agricultural practices were informed by public clocks. For example, she notes that "in order to understand the practical use of the dials, therefore, it is necessary to focus on the role of astrology in late medieval and Renaissance times".[3] As there is no single role of astrology over many centuries, a point forcefully made by Richard Dunn with respect to an English context, Addomine suggests that one might want to look carefully at medical astrology, and proceeds to explore that avenue, noting that providing a diagnosis or prognosis would be greatly impacted by small differences in time determinations.

One reason for addressing the Mantua clock in particular is that it seems to be the only example that shows planetary hours, which have an explicit association with astrological interpretation. This is an interesting point that merits more discussion. Perhaps, one might counter, this makes the very point that medieval or Renaissance Italian astronomical clocks did not generally have astrological import, if this is the only example with such obvious astrological content. Moreover, how would these planetary hours be viewed, used, or understood in a public setting? Would they be interpretable by members of the public? And by which members, all or a special subset? Or might they instead function in a rhetorical way, as a public reminder, perhaps of being observed by the divine, or of actions being predictable or traceable by one instructed in those arts? The latter is suggested by a text written just a year after the Mantua clock's construction in 1473. In this printed handbook, the author notes eight types of readings made possible by the clock, of which two are astrological. One example of the clock's potential use is highly suggestive: the handbook assures the reader that the clock could provide information that would identify a thief, his appearance, his home and even the location of stolen loot. Such a discussion, Addomine notes, is widely known in Islamic texts, yet another instance of demonstrating that we cannot rely solely on Ptolemaic principles for understanding medieval astrology and astrological instruments. In the event,

3 Marisa Addomine, *infra*, p. 139.

Addomine's examination of astrological pursuits associated with the Mantua astronomical clock illustrates the necessity of investigating specific contexts, and deserves a deeper inspection.

A more common instrument, the globe, will now also attract attention for potential astrological connections. The ubiquity of globes in portraits and in depictions of travellers, naval officers, natural philosophers and other well-educated gentlemen in the sixteenth through eighteenth centuries has previously yielded an awareness of the importance of astronomy in that period beyond mere professional engagement. Moreover, the titles of some of these depictions also allude to astrology. Typically, terrestrial and celestial globes were sold in pairs, along with instruction manuals for carrying out calculations enabled by the use of the globe. In their paper, Jim Bennett and Sylvia Sumira consider the matter of whether celestial globes were actually used to conduct astrological computations; more specifically, they consider whether globes could be used, and how.

Beyond finding images, which are suggestive but not definitive, they have located a letter by astrologer William Lilly in which he describes precisely how he has used his celestial globe to carry out an astrological computation. Just as one swallow does not a spring make, one letter does not a practice make, so they show how Joseph Moxon's *Tutor to Astronomie and Geographie* (1659) describes the process of using a celestial globe to divide the heavens into twelve houses, and outline a method described by Thomas Blundeville in his *Exercises* (1594), which includes the practical suggestion of inserting wooden wedges to prevent the globe from rotating inadvertently, a detail they describe as giving "some reassurance that he is describing a real operation".[4]

They conclude that it seems reasonable that ordinary globes could have been used for astrological purposes, and likely were, given the level of detail provided in these and other texts. They have also identified a couple of globes, as well as a painting, that include a semicircle of position, a globe accessory whose purpose was almost certainly astrological in function. Another piece of material evidence would come from globes with "signs of having been marked with planets or lunar nodes".[5] So far, no such marks on globes have been noted, and with her extensive experience in globe restoration, Sumira is certainly well situated to have found such already or to notice such in the future. This careful and cautious analysis gives explicit features to look for when we examine globes in collections. We look forward to updates on this topic.

4 Jim Bennett and Silvia Sumira, *infra*, p. 220.

5 *Ibid.*, p. 221.

Louise Devoy's investigation also directs our attention to a specific instrument, or rather, an element found on many instruments: the aspectarium. Her survey of over sixty devices, primarily in London's Science Museum, the British Museum, the Museum of the History of Science, Oxford, and Royal Museums Greenwich shows the importance of conducting surveys of actual instruments in order to develop an understanding of their purpose and function. Her clearly laid groundwork enables the discernment of a few trends. First, the inclusion of aspectaria coincides with the period of highest production for the instruments on which they appear, from the 1480s to the 1650s. Because this coincidence may be a selection effect, the inference is tenuous but nonetheless quite intriguing. How consistent are the glyphs over this large time period? What might we learn from that consistency or from those changes? One big payoff comes from her second inference, that German instruments differ distinctly from those made in London. The aspectaria on the former appear to be more aesthetic embellishment, often part of a lunar volvelle for converting between lunar and solar hours; by contrast, London makers designed their aspectaria with a ray diagram located within a zodiac circle necessary for carrying out astrological calculations. Devoy has provided valuable, suggestive background material for why this might be the case, and specific details to examine on aspectaria for gathering additional data about this noteworthy distinction.

These visible distinctions lead us to Richard Dunn's nicely delineated exploration of visual and literary imagery that depicts instruments in English astrological settings from 1600 to 1750. Dunn further restricts attention to matters of judicial astrology, the determination of an individual's character, of the propriety of a specific event, or even of the context or resolution of a question based on heavenly positions at the moment of its posing. That determination was aided by the construction of a chart or figure representing planetary positions and astrological parameters for interpretation. An instrument could enable observations and summarize or assist with the computation of observational and tabular data so that one could construct the astrological figure prior to or subsequent to an event. Dunn has identified diverse images featuring one or more instruments that could carry out this task, in whole or in part: a globe, an armillary, a cross-staff and dividers, as well as books and telescopes. The issue at hand, as Dunn and others in this volume note, is whether pictorial representations of these instruments point to genuine practices. Dunn adds to the usual list of pictorial sources by examining books of trades as well as other sources of astrological imagery, and concludes that illustrations continue into the eighteenth century to draw on established iconography, depicting types rather specific individuals whose actual practices can be confirmed.

Analysis of the casebooks of English astrological consultants provides considerable insight about the kinds of questions posed, as well as about the enormous volume of queries addressed. It strongly indicates that the development of the astrological figure was the central task of a consultation, aided by an ephemeris listing the astronomical data needed. And while an instrument could be useful, the consultants themselves say little about the use of instruments, with Dunn noting how one (Simon Forman) commented negatively about their usefulness, and how another (Richard Napier) merely notes the use of a sundial for finding the time. Nor do early seventeenth-century descriptions by John Melton and Thomas Tomkis help, as instruments in their accounts serve to legitimize the astrologer's work, to befuddle their clients, or even to satirize the astrologer.

For Dunn, as for other authors, the information at hand still leaves open the question about whether astrologers (or laypersons) actually used "scientific instruments" at all, much less routinely, in carrying out their astrological tasks. He suggests that we might think of use in a different way, analogous to the visual and literary roles played by instruments in representations. Rather than being used functionally, *qua* instrument, Dunn points to actual scientific instruments having a symbolic and didactic purpose. Thus, for example, it was not enough to have an illustration of an instrument; an actual, physical device was present in the consulting room to convey the authority of the astrologer in a way that a mere representation could not. Here, too, caution is needed. Images admit of rapidly changing styles and interpretive elements, as illustrated by the transition in just a few decades from symbolic representations in Mannerism to more naturalistic depictions by Caravaggio and his followers. How much more must we be cautious in inferring patterns drawn from images scattered over a century or more.

This blurriness about the nature and purpose of scientific instruments is further articulated in three other papers. Josefina Rodríguez-Arribas raises these issues prompted by her investigation of four sixteenth-century German plates, two in metal, two in paper. After examining their details, and showing how they might have been used, she concludes that the paper examples might have been models for "actual" metal instruments, and wonders whether or not these actually are instruments, much less *scientific* instruments. She notes that the two metal plates were likely designed for medicine and astrology, and represent practices and theories in place by the Middle Ages. While quite plausible, we do well to exercise caution in answering questions about the incorporation of devices into early modern practices by suggesting a connection to earlier practices whose incorporation of devices are even less well-established.

Stephen Johnston, in his analysis of two astrological instruments made by Thomas Hood that were "clearly intended as objects of both intellectual and visual consumption",[6] richly illustrates issues underlying the notion of blurriness. In his account, the interplay of manuscripts and printed texts, mathematical tables, public lecturing, astrological and navigational charts, and paper, wood and brass instruments for diverse purposes connects an array of materials whose descriptions, purposes and associated practices straddle both sides of any line demarcating the boundaries of science and, by extension, any associated definition of a scientific instrument.

Seb Falk's detailed description of the similarities and variations in astrolabic texts and devices emphasizes the fluidity of the very categories of text and instrument. In particular, he questions the notion of a finished instrument, pointing out that users often altered, adjusted, customized and recalibrated their devices. That flexibility, he proposes, emerges from a variety of desired movements from theoretical text to practical instrumentation; that transition might involve trading off speed versus accuracy, affordability versus durability, or computation versus presentation, to mention only a few examples. In fact, he notes, a text itself might offer sufficient instrumentality so as to preclude the need for a three-dimensional device.

The papers in this volume illustrate that we still know precious little about the use of real physical devices in actual astrological practice. That is itself an important result. We are left with many questions; in fact, more questions than when we began. Questions about *if*, not just *how*, the objects discussed were actually used in the service of astrology. Descriptions of highly variable astrological practices over time, from location to location, and in different political, theological, civic and social contexts should temper our expectations that instruments might play a consistent or predictable role. The examinations of instruments themselves reveal a diversity of parameters tracked and methods of computation, with documentation indicating that underlying concepts and actual practices can change even over the course of a given practitioner's career. In short, this volume shows that in light of astrological practice itself being highly idiosyncratic, our accounts of scientific instruments and their usage in the service of astrological practice should reflect that diversity accordingly.

Indeed, the current situation urges that we also consider an array of related, underlying issues. We have to some extent, perhaps inadvertently, entered into these investigations with implicit and explicit assumptions about the category we call "scientific instrument" and about what it means to use one. Put another way, given the variabilities and multiplicities of astrological practice,

6 Johnston, *infra*, p. 82.

in conjunction with the constancy and range of astronomical cycles, what is about those *practices* in virtue of which we might expect both regularity and diversity in the tools used in support of them? How do those astrological practices and astronomical cycles yield what we consider *scientific* instruments? Dunn and Rodríguez-Arribas in particular have moved in this direction, suggesting that we consider these devices and representations as visual tools. What is meant by a visual tool, and how a visual tool functions, in the context of astrological practice is a matter about which art historians, anthropologists and others can make valuable contributions to the current discussion.

Some decades ago, histories of science consisted of the histories of ideas, tracing the emergence in time of timeless truths to be contemplated. Later histories incorporated the contingent, historical circumstances that yielded universal, scientific principles, leading to more contingent, sociologically-sensitive histories in which the universality of science came into question. In this evolution, scientific practice and the roles of instrumentation have played more prominent roles. Whatever historiographical approach taken so far, and whatever historiographies are still to come, the history of science seems particularly sensitive to the tension between the particular and the universal, between the contextual locus characteristic of every historical event and the general principles of nature marking the *telos* of many scientific activities. Indirectly and directly, the papers in this volume point to the need to develop a more sophisticated historiography of scientific instruments, in which we come to a richer understanding of the changing and contingent roles played by the material culture of science.

Glossary

Richard Dunn

Astrology is a complex art with its own vocabulary and a number of different traditions. These have changed over time. This glossary covers the key astrological and related terms mentioned in this volume. It is intended primarily to give sufficient explanation to aid understanding of the individual chapters. For those wishing to explore the terminology further, the major sources consulted are listed at the end of the glossary.

accidental dignity see *dignity*.

almanac a book or set of tables, usually containing a calendar, with astronomical data and calculations, ecclesiastical and other anniversaries and other information, including astrological and meteorological forecasts.

almucantar a circle on the *celestial sphere* parallel to the *horizon*; a parallel of altitude.

almuten the most influential *planet* (as determined by the *dignities* of each) in a *chart*.

application / applying see *aspect*.

ascendant the degree of the *zodiac* rising over the eastern *horizon* at the time for which an astrological *figure* or *chart* is cast; often also applied to the whole *sign* rising.

aspect an astrologically significant angular relationship (measured from the Earth's centre) between two or more *planets* or *nodal points* in a *chart* (see *Table G1*). Each aspect operates fully when the angle is exact, but has a partial influence for as long as part of the *orb* of each planet maintains the appropriate angle. In this case, the faster moving planet is said to be applying as it approaches the aspect with the other planet or nodal point, separating as it moves away. Traditionally, an aspect is more powerful during application than separation. Ptolemaic astrology used five aspects, to which Johannes Kepler (1571–1630) added a further eight.

TABLE G1 Planetary aspects; based on Nicholas Campion, *The Practical Astrologer*, Twickenham, Hamlyn Publishing, 1987, p. 70, and Deborah Houlding, "Glossary of terms", www.skyscript.co.uk/gl/aspect.html (accessed 13 November 2015)

Name	Symbol	Angle (degrees)	Strength	Nature
Ptolemaic aspects				
Conjunction	☌	0	Major	neutral
Sextile	⚹	60	Major	beneficial
Square or quartile	□	90	Major	harmful
Trine	△	120	Major	beneficial
Opposition	☍	180	Major	harmful
Keplerian aspects				
Semi-sextile	⚺	30	Minor	beneficial
Decile	⊥	36	Minor	beneficial
Octile	∠	45	Minor	harmful
Quintile	Q	72	Minor	beneficial
Tridecile	Td	108	Minor	beneficial
Sesquiquadrate or sesquiquartile or trioctile	⚼	135	Minor	harmful
Biquintile	bQ	144	Minor	beneficial
Quincunx	⚻	150	Minor	harmful

aspectarium an aide-mémoire depicting the *aspects* between the planets.
astronomical hours see *equal hours.*
azimene (or **lame**) see *degrees.*
Babylonian hours see *equal hours.*
Bohemian hours see *equal hours.*
Caput Draconis or **head of the dragon** the north (or ascending) node of the Moon, where it crosses the *ecliptic*; the most powerful *nodal point* in the *chart*, favourable for affairs governed by the *house* in which it is located; see also *Cauda Draconis.*
cardinal houses / signs or cardines the *houses* (first, fourth, seventh and tenth) or *signs* (Aries, Cancer, Libra, Capricorn) falling on the cardinal angles.
casting the rays According to ancient Greek astrology each *planet* casts seven rays of influence to other points of the *ecliptic*. The rays are defined by reference to a hexagon, square and equilateral triangle. In the simplest method, the

polygons are inscribed in the ecliptic with one apex at the planet's position. For interpretation, the theory of rays and *aspects* between the planets was often related to the theory of the aspects between the *signs*. Thus the trine ray or aspect was regarded as beneficial, the sextile as less beneficial, the quartile as harmful, and opinions differed as to the qualities of the opposite ray and aspect. Medieval Islamic astrologers enthusiastically adopted the theory of the seven rays.

Cauda Draconis **or tail of the dragon** the south (or descending) node of the Moon, where it crosses the *ecliptic*; regarded as an unfavourable influence for affairs governed by the *house* in which it is located; see also *Caput Draconis*.

celestial sphere an imaginary sphere, with Earth at its centre, on which all other celestial objects are considered to lie.

chart the astrological diagram drawn for a specific time (*Fig. G1*), showing the positions in the heavens of the *planets, luminaries, nodal points* and *cusps* of the *houses*; synonymous with *figure* and, in modern usage, *horoscope*. See also *division, house*.

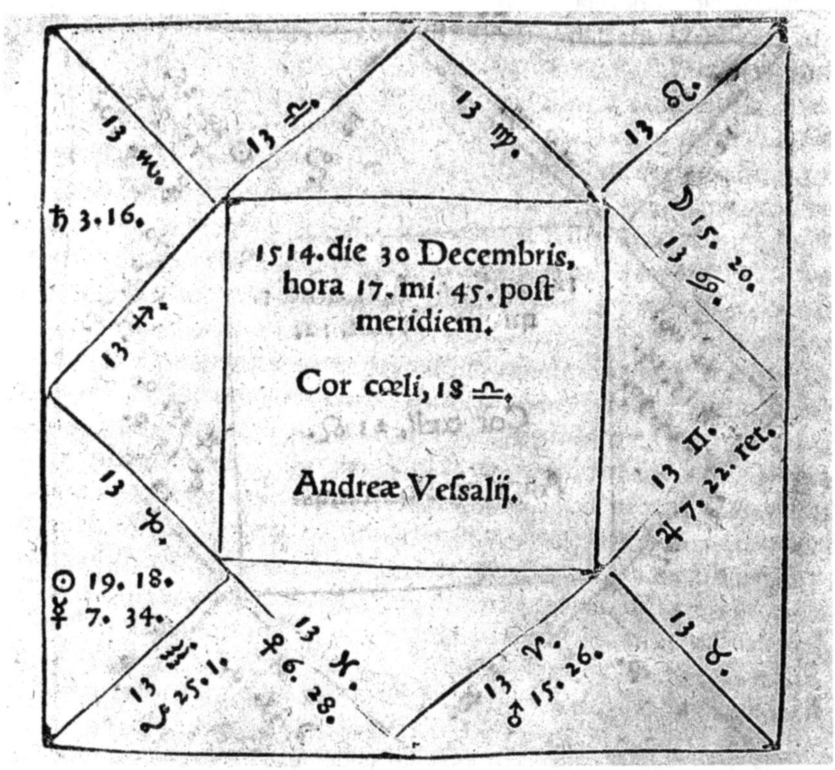

FIGURE G1 Astrological chart for the birth of Andreas Vesalius on 30 December 1514; from Girolamo Cardano, *Libelli quinque*, Nuremberg, Joannes Petreius, 1547, fol. 178 (Wellcome Library, London).

chronocrator a transliteration of a Greek term used to refer to a planet activated in a *chart* for a certain timespan; literally meaning "lord of time" because the planet is activated as a *lord* or *ruler* in the *chart* for a specific period.

conjunction see *aspects*.

crisis / critical day the theory of critical days links the course of an illness to the Moon's phases, with crises or critical days at each quarter of the Moon's cycle (i.e. every seven days), at which time the patient might improve or get worse.

cusp the beginning or entrance of a *house* or *sign*; see also *division*.

debility see *dignity*.

decan a 10° division of each *sign* into first, second and third decan, each ruled by a specific *planet*; also called face.

degrees individual degrees of the *zodiac* may have particular characteristics that come into play when any of the *planets* or *nodal points* are found there. In particular, they may influence the physical or mental characteristics of the subject of a *nativity* if the *ascendant* or its *lord*, or possibly the Moon, is found in one of them. They can have a similar interpretation in *horary astrology*. The most common characteristics are:

- masculine or feminine—helpful in determining the sex of a thief or unborn child, for example;
- lucid (or light), dark, smoky or empty (or void);
- pitted (or deep), indicating either something of the subject's complexion or that a *querent* may need help if they are to improve their condition;
- lame (or deficient or azimene)—perhaps referring to the subject, but also regarded as debilitating when on the *ascendant* or otherwise prominent in the *chart* (e.g. with the Moon in that degree).
- increasing fortune—indicating that the querent will gain riches.

depression a planet is said to be in depression when it is in the *sign* or degree opposite its *exaltation*; also called fall; see also *debility*, *dignity*.

descendant strictly speaking, the *cusp* of the seventh *house* (i.e. the degree of the *zodiac* setting below the western *horizon*), but sometimes also used to refer to the whole seventh house.

detriment a *planet* is in detriment when it is in the *sign* opposite its own *domicile*.

dignity the situation of a *planet* in which its influence is heightened due to its position in the *zodiac*, the *aspects* it has with other planets or *nodal points*, or its relationship to the *houses*. Essential dignities occur when a planet is:

- in its *domicile* or in *mutual reception* by domicile;
- in *exaltation* or in *mutual reception* by exaltation;
- in its own *triplicity*;
- in its own *term*;
- in its own *face*.

Accidental dignities derive from other relationships (e.g. with the *houses*). Essential and accidental dignities may come into play at the same time. The opposite of dignity is debility, a relationship that in general weakens the influence of the planet. The essential debilities include a planet being in *detriment* and in *depression*.

direction the study of future configurations of the *planets* and angles in relation to a radical *chart* (i.e. a chart at a key moment such as a person's birth or the beginning of an event or action) in order to understand the future course of events; also known as *progression*.

division the method of defining the *cusps* or boundaries of the *houses* in a *chart*. Many different methods have been devised over the centuries, often named after practitioners. The more important methods include:

- equal-house (sometimes called Ptolemaic) method—the *ecliptic* is divided into twelve equal divisions of 30°, starting from the degree of the *ascendant*, which acts as the *cusp* of the first house.
- Regiomontanus or equatorial fixed-boundary method, also called rational—the celestial equator is divided equally into twelve, with the divisions projected onto the ecliptic along great circles that take in the north and south points on the *horizon*; each house spans 30°. Though attributed to Regiomontanus (Hans Müller of Königsberg, 1436–1476), the method can be traced back to at least the eleventh century and was the main European method for several centuries, being employed by prominent seventeenth-century astrologers including William Lilly (1602–1681).
- Campanus method—the *prime vertical* is divided equally into twelve, with the divisions projected onto the ecliptic along great circles that take in the north and south points on the horizon. Although attributed to Johannes Campanus (*c.* 1220–1296), the method is known to have been used by Al-Bīrūnī (b. 973).
- Alcabitius, common or standard method—the parallel of declination on which the ascendant is located is divided equally from the ascendant, using lines of constant right ascension to project the house cusps onto the ecliptic.
- Porphyry method—each quadrant of the ecliptic is divided into three equal parts. This is the oldest system of quadrant-style house division. Although later attributed to Porphyry of Tyre (*c.* 234–*c.* 305), it had already been described by the second-century astrologer Vettius Valens (120–*c.* 175).
- Placidus or hour-line method—the paths drawn for each degree of the ecliptic to move from the *Imum Coeli* to the *horizon*, and from the horizon to the *Medium Coeli*, are trisected to define the cusps of houses 2, 3, 11 and 12. The cusps of houses 8, 9, 5 and 6 are opposite these. The system was named after the Italian Benedictine monk, Placidus de Titis (Placido de Titi, 1603–1668),

who popularized its use during the seventeenth century, although it had been described earlier.

domicile or mansion the *sign* ruled by a specific *planet* (see *Table G3*). When a planet is in its domicile, which is an essential *dignity*, its influence is enhanced; see also *dignity, sign*.

ecliptic the great circle that is the apparent path of the Sun among the constellations in the course of a year.

election / electional astrology a form of casting and interpreting a *chart* in order to determine the best time for commencing a particular action or endeavour; for example, marriage, a journey or a law suit.

ephemeris / ephemerides a set of tables listing the positions of the *planets* and *nodal points* and other astronomical data.

equal hours or **equinoctial hours** the time system by which each whole day is divided into 24 equal parts. Typically, the day was divided either into 24 equal hours beginning at midnight or two twelve-hour periods beginning at midnight and noon. In the first case the hours were referred to as "large" or "great" hours; in the latter case they were "small", "common", "ordinary", "German" or "French" hours. Three additional systems based on "large" hours were common in the Renaissance and early modern period: "Italian", "Bohemian" or "foreign" hours, reckoned from sunset; "Babylonian" or "Greek" hours, reckoned from sunrise; and "Nuremberg" hours, which counted daylight hours by the Babylonian system and night hours by the Italian system. The system of "astronomical hours", which was also based on equal hours, reckoned the day from noon to noon. See also *unequal hours*.

equatorium an instrument for finding the positions of the *planets* using a geometrical model to represent the position of the bodies in question.

essential dignity see *dignity*.

exaltation a specific degree, or sometimes a *sign*, of the *zodiac* in which a *planet* exerts its most powerful influence. The opposite of exaltation is fall or *depression*. See also *dignity*.

face see *decan*.

fall see *depression*.

figure a synonym for *chart*.

French hours see *equal hours*.

genethlialogy the practice of calculating and making predictions from a *chart* for the moment of birth.

geniture the *chart* drawn for a person based on his/her time of birth; also called a *nativity* or *horoscope*.

German hours see *equal hours*.

head of the dragon see *Caput Draconis*.

hebdomada a seven-day cycle in which each day corresponds to one of the seven *planets*, with the first day being Saturday (Saturn).

horary / horary astrology the art of making a forecast regarding a specific question from a *chart* or *figure* drawn for the moment the question is asked.

horizon the circle around the Earth separating the hemisphere visible to the observer (or *querent*) from the hemisphere that is invisible. The eastern horizon is the degree rising in the east.

horoscope until the eighteenth century, horoscope referred to the *ascendant* degree in the chart; in modern usage, it is synonymous with *chart*.

hours see *equal hours, unequal hours*.

house the *division* of the sky into a number of regions (typically twelve, although some systems use other numbers) for the purpose of creating a *chart*, usually beginning with the first house at the rising *horizon*. Each house governs a different area of life (*Fig. G2*).

iatromathematics medical astrology.

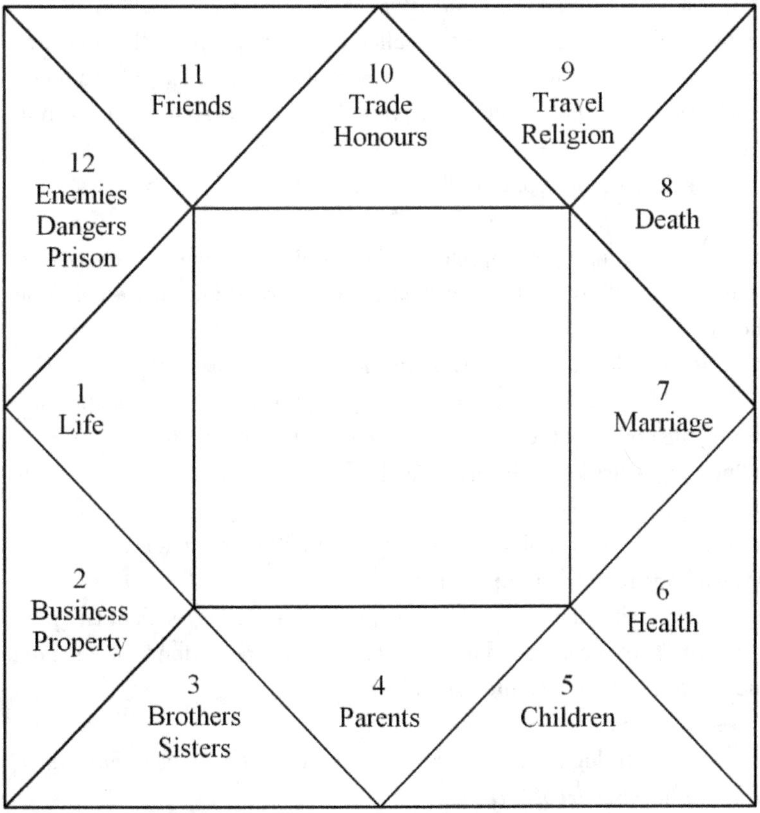

FIGURE G2 The houses of the traditional chart showing the areas of life they govern.

Imum Coeli the point opposite the *Medium Coeli* (mid-heaven), also known as the anti-mid-heaven; usually the *cusp* of the fourth *house.*

Italian hours see *equal hours.*

judicial astrology a form of astrology seeking to make predictions about small-scale phenomena, particularly including the futures of individuals based on their *geniture.*

light a term for the Sun and Moon; the term luminary can also be used.

limits see *terms.*

lord properly applied, the most powerful *planet* in a *chart.*

lord of time the *planet* ruling a particular hour of the day; also called *chronocrator* or planetary ruler; see also *planetary hours.*

lord of the triplicity see *tripicity.*

lower mid-heaven a variant term for *Imum Coeli.*

luminary see *light.*

lunar mansion a division of the path of the Moon into equal arcs, often 27 or 28 segments, each containing a specific asterism (group of stars). Each mansion has a name and specific interpretation, affecting how a *planet* in that mansion exerts its influence. Lunar mansions were used in the Islamic world as a merger of pre-Islamic and Indian systems, and were then introduced into Europe during the Middle Ages.

lunar nodes see *Cauda Draconis* and *Caput Draconis.*

mansion see *domicile.*

Medium Coeli a term properly applied to the culminating degree of the *ecliptic* (the point of the ecliptic on the *meridian*); also known as mid-heaven or upper mid-heaven.

melothesia or **melothesic man** see *Zodiac Man.*

meridian the plane containing the observer/*querent* and both celestial poles, or containing the observer, the *zenith* and a pole; more colloquially, the north-south line at a particular position on Earth.

mid-heaven see *Medium Coeli.*

mundane astrology a form of astrological prediction limited to forecasts of trends over large regions or populations.

mutual reception in principle, mutual reception occurs when each of two *planets* is in a *dignity* of the other. In practice, only mutual reception between *domicile* and *exaltation* are considered significant.

nativity see *geniture.*

nodal point any astrologically sensitive degree on the *ecliptic*; see *Caput Draconis, Cauda Draconis, Part of Fortune.*

Nuremberg hours see *equal hours.*

octile see *aspect*.

opposition see *aspect*.

orb or ***orbes planetarum*** the area of influence within which *planets* in *aspect* are said to exert an influence while applying or separating. In some cases, the aspect itself can be ascribed an orb.

Part of Fortune a *nodal point* in the *chart* that is the same distance of zodiacal arc from the *ascendant* as the Moon is from the Sun (according to one of the methods for calculating it), marking the point where the Moon would be at the time of symbolic sunrise; held to be beneficial for the affairs governed by the *house* in which it is situated.

planetary hours each hour of the day is said to be ruled by one of the *planets* (*Table G2*), a cycle that gives rise to the names of the days of the week (from the planet ruling the first hour). This sequence is noticeable in the names of the days of the week in Romanesque languages; for example, *Martedì* (Tuesday) from *Marte* (Mars), *Mercoledì* (Wednesday) from *Mercurio* (Mercury), *Giovedì* (Thursday) from *Giove* (Jupiter) and *Venerdì* (Friday) from *Venere* (Venus) in Italian. It is also apparent in some English names (notably Saturday, Sunday and Monday), but less so in others. Friday, for instance, comes from the Old English Frīġedæġ (day of Frigg), reflecting the association of the English goddess Frigg with the Roman goddess Venus.

TABLE G2 The planetary hours of the day and night

Planetary hours of the day

Hour	Sunday	Monday	Tuesday	Wednesday	Thursday	Friday	Saturday
1	Sun	Moon	Mars	Mercury	Jupiter	Venus	Saturn
2	Venus	Saturn	Sun	Moon	Mars	Mercury	Jupiter
3	Mercury	Jupiter	Venus	Saturn	Sun	Moon	Mars
4	Moon	Mars	Mercury	Jupiter	Venus	Saturn	Sun
5	Saturn	Sun	Moon	Mars	Mercury	Jupiter	Venus
6	Jupiter	Venus	Saturn	Sun	Moon	Mars	Mercury
7	Mars	Mercury	Jupiter	Venus	Saturn	Sun	Moon
8	Sun	Moon	Mars	Mercury	Jupiter	Venus	Saturn
9	Venus	Saturn	Sun	Moon	Mars	Mercury	Jupiter
10	Mercury	Jupiter	Venus	Saturn	Sun	Moon	Mars
11	Moon	Mars	Mercury	Jupiter	Venus	Saturn	Sun
12	Saturn	Sun	Moon	Mars	Mercury	Jupiter	Venus

TABLE G2 The planetary hours of the day and night (*cont.*)

Planetary hours of the night

Hours	Sunday	Monday	Tuesday	Wednesday	Thursday	Friday	Saturday
1	Jupiter	Venus	Saturn	Sun	Moon	Mars	Mercury
2	Mars	Mercury	Jupiter	Venus	Saturn	Sun	Moon
3	Sun	Moon	Mars	Mercury	Jupiter	Venus	Saturn
4	Venus	Saturn	Sun	Moon	Mars	Mercury	Jupiter
5	Mercury	Jupiter	Venus	Saturn	Sun	Moon	Mars
6	Moon	Mars	Mercury	Jupiter	Venus	Saturn	Sun
7	Saturn	Sun	Moon	Mars	Mercury	Jupiter	Venus
8	Jupiter	Venus	Saturn	Sun	Moon	Mars	Mercury
9	Mars	Mercury	Jupiter	Venus	Saturn	Sun	Moon
10	Sun	Moon	Mars	Mercury	Jupiter	Venus	Saturn
11	Venus	Saturn	Sun	Moon	Mars	Mercury	Jupiter
12	Mercury	Jupiter	Venus	Saturn	Sun	Moon	Mars

planetary ruler each *planet* is held to rule over a particular *sign* or signs (*Table G3*).

TABLE G3 The planetary rulers according to Ptolemaic astrology

	Day	Night
Sun	Leo	–
Moon	Cancer	–
Mercury	Virgo	Gemini
Venus	Libra	Taurus
Mars	Scorpio	Aries
Jupiter	Sagittarius	Pisces
Saturn	Capricorn	Aquarius

planets for traditional astrological purposes (and for all purposes until the eighteenth century), the seven celestial bodies seen to be moving against the background of the fixed stars: Sun (☉); Moon (☽); Mercury (☿); Venus (♀); Mars (♂); Jupiter (♃); Saturn (♄). Each planet exerts a distinctive influence.

prime vertical the great circle taking in the *zenith* and east and west points on the observer's or *querent's horizon.*

progression see *direction.*

promittor that which promises fulfilment, generally meaning a *planet* that will form a particular *aspect* through *progression* or *direction.*

quality this can mean the qualities (hot, cold, humid and dry) of the four elements. Thus air is hot and humid, earth is cold and dry, fire is hot and dry, and water is cold and humid. It can also refer to the three qualities (cardinality, fixity and mutability), by means of which each of the four elements finds expression in the *signs* of the *zodiac*. Aries, Cancer and Libra, for example, are of the *cardinal* quality.

quartile see *aspect.*

querent the person who asks a question in *horary astrology.*

ray used in traditional texts as a synonym for influence; see also *casting the rays.*

retrograde used to describe a *planet* when it appears (from Earth) to be moving backwards in the *zodiac*, generally held to have a negative connotation for the planetary influence.

revolution(s) of the years an astrological technique in which the *chart* is progressed to the future point or points at which the Sun has returned to the same place in the *zodiac* (e.g. the place it was first in at the time of a person's birth), used for assessing issues that will arise over the period.

ruler see *planetary ruler*

seasonal hours see *unequal hours.*

separation / separating see *aspect.*

sextile see *aspect.*

sign one of the 30° arcs of the *zodiac*, through one of which the Sun passes each month. The signs are named after the twelve constellations with which they coincided at a former epoch. The signs are grouped into four groups called *triplicities* or trigons, each of which is associated with one of the four elements (fire, air, water, earth; see *Table G4*).

Spanish hours see *equal hours.*

square see *aspect.*

tail of the dragon see *Cauda Draconis.*

term irregular divisions of the *signs*, generally into five segments, each falling under the influence of one of Mercury, Venus, Mars, Jupiter or Saturn; also known as *limits*. Girolamo Cardano (1501–1576) devised a system of seven terms, each ruled by one of the seven *planets.*

trigon see *triplicity.*

trine see *aspect.*

triplicity or **trigon** a group of three *signs*, each separated by 120°. Each triplicity is linked with one of the elements, has a planetary ruler (or lord) by day, another by night, and a third known as a companion (*Table G4*).

TABLE G4 The triplicities and their rulers

Element	Signs	Ruler by day	Ruler by night	Companion
air	Aquarius; Libra; Gemini	Saturn	Mercury	Jupiter
earth	Capricorn; Virgo; Taurus	Venus	Moon	Mars
fire	Sagittarius; Leo; Aries	Sun	Jupiter	Saturn
water	Pisces; Scorpio; Cancer	Venus	Mars	Moon

unequal hours the system in which the periods of daylight and darkness are each divided into twelve equal intervals, with the hours counted from sunrise and sunset. This was the most common system used for counting hours in medieval and Renaissance Europe. Noon is defined as the moment when the Sun reaches its greatest altitude and marks the midpoint of the daylight period; sunrise and sunset occur six hours earlier and later. The lengths of each daytime hour and night-time hour are unequal (except at the spring and autumn equinoxes and on the equator) and vary according to geographical latitude and date. Unequal hours can also be known as temporal, seasonal, Jewish or planetary hours. See also *equal hours*.

upper mid-heaven a variant term for *Medium Coeli.*

volvelle a device that rotates, usually referring to one or more discs which turn within a circular scale.

zenith the point on the *celestial sphere* directly above the observer or *querent.*

zij an astronomical text with tables compiled in the Islamic world.

zodiac the belt of the *celestial sphere* extending about 8 or 9° on each side of the *ecliptic*, within which the apparent motions of the *planets* take place. The zodiac is usually divided into twelve equal *signs.*

Zodiac Man the image of a human linked with the twelve zodiacal *signs*, indicating which signs rule which parts of the body (*Fig. G3*); also called melothesia or melothesic man.

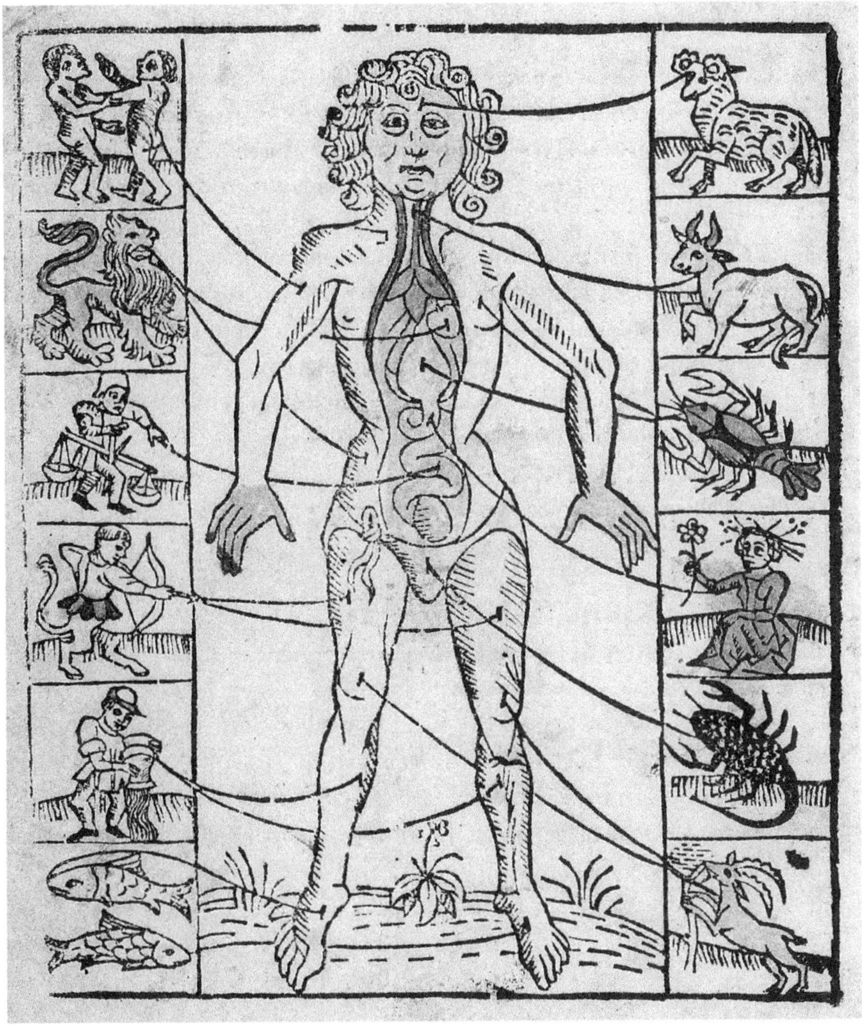

FIGURE G3 Zodiac Man; from *Kalendar Deutsch*, Augsburg, J. Blaubiren, *c.* 1483 (Wellcome Library, London).

Sources

- Silke Ackermann, "Astrological Scales on the National Maritime Museum Astrolabes", in Koenraad Van Cleempoel, *Astrolabes at Greenwich*, Oxford and London, Oxford University Press/National Maritime Museum, 2005, pp. 73–89.
- *The Astrology Dictionary*, http://theastrologydictionary.com/ (accessed 23 March 2017).
- Nicholas Campion, *The Practical Astrologer*, Twickenham, Hamlyn Publishing, 1987.
- J. Casulleras, J.P. Hogendijk, "Progressions, Rays and Houses in Medieval Islamic Astrology: A Mathematical Classification", *Suhayl* 11 (2012), pp. 33–102.
- *EPACT*, www.mhs.ox.ac.uk/epact (accessed 15 June 2017).
- Deborah Houlding, "Glossary of Traditional Astrological Terms", *Skyscript*, www.skyscript.co.uk/gl/glossary.html (accessed 15 June 2017).
- Deborah Houlding, "The Problems of House Division", *Skyscript*, www.skyscript.co.uk/houprob_print.html (accessed 15 June 2017).
- Fred Gettings, *Dictionary of Astrology*, London, Routledge & Kegan Paul, 1985.
- Annabella Kitson (ed.), *History and Astrology: Clio and Urania Confer*, London, Unwin Hyman, 1989, pp. 253–257.
- *Medieval Astrology Guide*, www.medievalastrologyguide.com (accessed 25 September 2017).
- John D. North, *Horoscopes and History*, London, Warburg Institute, 1986.
- *OED Online*, www.oed.com (accessed 23 March 2017).
- *Renaissance Astrology*, www.renaissanceastrology.com (accessed 22 June 2017).
- James Wilson, *A Complete Dictionary of Astrology*, London, William Hughes, 1819.

Index

For Arabic names:

- 'al' has not been counted for indexing, but 'Abu', 'Ibn', etc. have been. Thus al-Bīrūnī appears under under 'b', Abū Ma'shar appears under 'a'
- 'b.' has been written as 'Ibn'